BURLEIGH DODDS SCIENCE: INSTANT INSIGHTS

NUMBER 63

Proximal sensors in agriculture

T0200170

burleigh dodds
SCIENCE PUBLISHING

Published by Burleigh Dodds Science Publishing Limited
82 High Street, Sawston, Cambridge CB22 3HJ, UK
www.bdspublishing.com

Burleigh Dodds Science Publishing, 1518 Walnut Street, Suite 900, Philadelphia, PA 19102-3406, USA

First published 2023 by Burleigh Dodds Science Publishing Limited
© Burleigh Dodds Science Publishing, 2023, except the following: Chapter 3 was prepared by a
U.S. Department of Agriculture employee as part of their official duties and is therefore in the public
domain. All rights reserved.

British Library Cataloguing in Publication Data
A catalogue record for this book is available from the British Library

ISBN 978-1-80146-423-9 (Print)
ISBN 978-1-80146-424-6 (ePub)

DOI: 10.19103/9781801464246

Typeset by Deanta Global Publishing Services, Dublin, Ireland

Contents

4   Using remote and proximal sensor data in precision agriculture
     applications   77
     *Luciano S. Shiratsuchi and Franciele M. Carneiro, Louisiana
     State University, USA; Francielle M. Ferreira, São Paulo State
     University (UNESP), Brazil; Phillip Lanza and Fagner A. Rontani,
     Louisiana State University, USA; Armando L. Brito Filho, São
     Paulo State University (UNESP), Brazil; Getúlio F. Seben Junior,
     State University of Mato Grosso (UNEMAT), Brazil; ny N. Brandao,
     Brazilian Agricultural Research Corporation (EMBRAPA), Brazil;
     Carlos A. Silva Junior, State University of Mato Grosso (UNEMAT),
     Brazil; Paulo E. Teodoro, Federal University of Mato Grosso do Sul
     (UFMS), Brazil; and Syam Dodla, Louisiana State University, USA*

Series list

Chapter 1

Proximal crop sensing

Richard B. Ferguson, University of Nebraska-Lincoln, USA

1 Introduction

2 The evolution of crop sensors

3 Current issues in sensor development

4 Case studies

5 Conclusion: sustainability and environmental implications

6 Future trends for research

7 Where to look for further information

8 References

1 Introduction

1.1 The importance of sensors

The use of sensors to evaluate crop status is, in one sense, ancient; farmers have manually evaluated plants and crop development from the beginning of crop cultivation. They could observe if individual plants appeared to be growing normally, and could begin to estimate yield potential when fruiting structures developed as plants matured. They learned to associate certain symptoms with specific stresses and could differentiate water deficit from nutrient deficiency and foliar disease from insect infestation. Farmers today are no different from their ancestors in this regard. However, one challenge for today's crop producer is scale. While a farmer a thousand years ago might manage a fraction of a hectare manually and observe his crops on a daily basis, farmers today may manage hundreds or even thousands of hectares, enabled by mechanization, herbicides, irrigation and other tools of modern agriculture. Consequently it is not possible for a farmer today to be physically in contact with crops during the growing season in the same way as their ancestors.

Today, farmers are increasingly reliant on sensors in their farming operations. All the equipment used in crop production - tractors, planters, combines and so forth - have a multitude of sensors to help optimize machine operation. Sensors are also becoming increasingly useful in monitoring crop and soil conditions. This chapter addresses the use of proximal sensors to evaluate a

http://dx.doi.org/10.19103/AS.2017.0032.02

crop during the growing season. The term 'proximal' indicates the sensor is in close proximity to the crop. Thus, its use is differentiated from other sensor applications such as remote sensing (using aerial or satellite sensor platforms) or soil sensing (using proximal or *in situ* sensors to evaluate soil properties, rather than crop properties). The term 'proximal' is fuzzy, as developments in unmanned aerial systems (UASs) have brought into the proximal realm what had been historically classified as remote sensing technologies. In this chapter, proximal crop sensing will refer to any sensor used to evaluate the properties of a crop, ranging from physical contact with the crop canopy to a few metres above the canopy.

1.2 Crop properties of interest

There are a variety of crop properties for which proximal sensing may be useful. From the perspective of the crop producer, properties that influence yield, or in some cases crop quality, and issues that can be detected and managed if necessary during the growing season are of greatest interest. Biomass accumulation, crops' water status, nutrient deficiency (particularly nitrogen), disease onset and weed and insect infestation are all factors that crop growers may wish to monitor throughout the season. For a number of these properties there are specific periods during the growing season where monitoring is particularly critical, depending on the crop and region in which it is grown. Proximal sensing, as opposed to remote sensing, has a potential advantage in that sensors on satellite and aerial platforms will be more influenced by weather and clouds, potentially limiting the utility of remote sensing at critical times in the growing season.

2 The evolution of crop sensors

2.1 Contact or in situ sensors

Crop sensors typically infer specific crop properties, rather than directly measuring them. There are some exceptions to this generalization, mostly sensors directly attached to or placed among plants. Sap flow sensors, for example, can be attached directly to a plant stem to estimate transpiration (Steinberg et al., 1989). Leaf area index (LAI) can be measured with ground-based sensors as well as with destructive methods (Bréda, 2003). However, measurement of sap flow and LAI with static instruments is time-consuming and does not provide spatial information, at least without intensive effort. A plant contact sensor which is a relatively simple biomass sensor developed for use with cereals is the Crop-meter (Ehlert and Dammer, 2006). This sensor, based on the pendulum principle, is mounted on a vehicle driven through the crop. The greater the biomass of the standing crop, the more the Crop-meter is deflected. For small-grain cereals, crop biomass at later growth stages can be a good predictor of grain yield potential, and thus, can be used for site-specific management of nitrogen fertilizer, growth regulators and fungicide in real time based on Crop-meter input.

2.2 Ranging sensors

Another sensor approach to estimate biomass and or crop height is range-finding, or distance measurement, using acoustic, laser or radar sensors. A review article by Dworak et al. (2011) cites a range of approaches using acoustic and electromagnetic (EM) wave ranging to assess crop stand, estimate biomass or evaluate canopy characteristics. Lee and Ehsani (2009), for example, demonstrated use of a laser scanner to quantify geometric characteristics of citrus trees, which can be useful for yield prediction, water consumption estimation, health monitoring and long-term productivity monitoring. Shiratsuchi (2011) demonstrated that integration of acoustic ranging sensors to estimate crop canopy height with multispectral and thermal sensors was useful for the detection of water stress in maize.

2.3 Electromagnetic (EM) sensors

The majority of crop canopy sensors in use or development today utilize the EM spectrum to characterize crop properties (Fig. 1). Such sensors can be passive

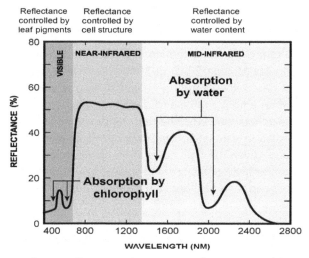

Figure 1 Primary factors influencing plant canopy reflectance in visible-to-mid-infrared wavelengths.

– using reflected, scattered or emitted energy – or active, using an internal pulsed or modulated energy source and measuring reflectance or fluorescence from that source. Radar, for example, is an active sensor. Most crop canopy sensors use reflectance, or in some cases transmission, in visible and near-infrared wavelengths to infer plant properties. The EM spectrum of interest ranges between the visible (400–700 nm), near-infrared (700–1300 nm), mid-infrared (1300–2500 nm) and thermal infrared regions (typically the principal region of interest is 8000–14000 nm). The primary factors influencing light absorption/reflectance in the visible portion of the spectrum are plant pigments, such as chlorophyll, carotenoids and anthocyanins. Cell structure, water content and canopy architecture primarily influence canopy reflectance in the near-infrared region, and water content in leaves influences reflectance in the mid-infrared region (Fig. 1).

In many cases, absolute reflectance in a given wavelength is not very informative about plant stress. Reflectance is related to various plant attributes, such as plant pigment concentrations, cellular structure, leaf water content or canopy architecture. The concept of a vegetation index (VI) allows the inference of specific plant properties and sources of stress. A VI uses the relationship of reflectance in two or more spectral regions, combined in an equation, to infer plant properties. More than 150 VIs have been described, with the most widely known being the normalized difference vegetation index (NDVI) (Rouse et al., 1974). The NDVI uses the equation below, resulting in values ranging from −1.0 to 1.0.

$$NDVI = (NIR - RED)/(NIR + RED) \qquad (1)$$

Wallihan (1973) described the development of a portable reflectance meter to estimate chlorophyll content of plant leaves. This can be thought of as a precursor to the chlorophyll meter, which is an instrument commonly used in plant research today. The most commonly used chlorophyll meter available today is the Konica Minolta SPAD 502 Plus (Konica Minolta, 2017) which measures light transmittance through a leaf surface at wavelengths of 650 and 940 nm when a light source and detector clamp are placed around a plant leaf. Researchers have documented the performance of this meter or preceding models in estimating leaf chlorophyll concentration (Markwell et al., 1995), and, given the high correlation between leaf chlorophyll and leaf nitrogen content, the capacity of the chlorophyll meter to detect N stress and predict the need for N fertilization (Piekielek and Fox, 1992; Follett et al., 1992; Blackmer and Schepers, 1995; Bundy and Andraski, 2004; Scharf et al., 2006). The SPAD-502 meter was adapted by Yara International ASA (Oslo, Norway) (N-Tester) with calibrations for N management of various crops. Ortuzar-Iragorri et al. (2005) found that normalization of N-Tester measurements with a non-N-limiting reference enabled the prediction of plant N concentration for soft red winter wheat. Varvel et al. (2007) described the concept of normalizing chlorophyll meter readings within a field for N fertilizer management of maize, using a non-limiting fertilizer N rate as a reference. This concept, termed a sufficiency index (SI), recognized that absolute chlorophyll meter readings may vary with cultivar, soil N supply and spatial distribution, within a field. In this concept, chlorophyll meter information is calibrated to a specific cultivar and field. Such calibration can also normalize for stresses other than N, such as sulphur, that may influence chlorophyll levels within leaves. Varvel et al. (2007) developed a fertilizer N recommendation algorithm for maize based on the SI concept.

Another type of handheld sensor uses principles of EM transmission as well as fluorescence to estimate leaf pigments and, consequently, nutrient stress or other abiotic stresses. The Dualex Scientific™ (Force-A, 2017) clips to a leaf and uses light-emitting diodes (LEDs) in ultraviolet (UV), visible and NIR wavelengths to estimate chlorophyll and polyphenol content. Polyphenols have been shown to be associated with plant stress factors such as N availability (Samborski et al., 2009). Tremblay et al. (2007) found that use of the Dualex in combination with a SPAD chlorophyll meter to calculate a ratio was more sensitive in detecting corn N status than using either instrument alone. Force-A company also manufactures portable handheld fluorometers measuring at UV and visible wavelengths (Multiplex Research™ and Multiplex 330™) intended for the assessment of abiotic stresses, in particular, disease detection (Sankaran and Ehsani, 2012).

While a contact sensor such as the chlorophyll meter may provide useful information about a given plant, it becomes laborious to use it to represent a group of plants or a field. Researchers have found the chlorophyll meter to be a useful proximal sensing tool in small-plot research, where readings from many plants within a plot can be averaged to represent effects of a treatment on chlorophyll levels. There are extension publications on the use of a chlorophyll meter for crop management (Shapiro et al., 2013); yet, the labour challenges associated with chlorophyll meter use preclude widespread adoption by crop producers for N fertilizer management.

2.4 Mobile EM sensors

Passive sensor systems

Among the first mobile EM sensors intended for commercial, field-scale evaluation of crop N status was the Yara N-Sensor™ (Yara International ASA, 2017). The N-Sensor™ is a passive, tractor-mounted spectrometer system consisting of two spectrometers, with one to scan the crop canopy to the side of the vehicle, and the other to measure ambient light to correct the reflected signal in real time in selected wavelengths between 450 and 900 nm (Zebarth et al., 2003; Samborski et al., 2009). Reflectance is used to calculate NDVI or other VIs of interest. The tractor-mounted system then uses an algorithm to calculate an optimal N fertilizer rate for the scanned region and relay this information to a controller to vary the application rate of N fertilizer on the go. The N-Sensor™ has been used primarily for spatial N management of small grains such as wheat (Berntsen et al., 2006; Jørgensen and Jørgensen, 2007), but has also been used for corn at early growth stages (Tremblay et al., 2009) and potato (Zebarth et al., 2003).

Active sensor systems

While a passive system can be potentially corrected for ambient light as with the N-Sensor™, there are still challenges with clouds, sun angle and early/ late time of day use. To precisely calculate reflectance, passive sensors need to measure incident light, to know the characteristics of the incoming light resource and to measure reflectance from the crop canopy at the same time (Raun et al., 2001). Consequently, active sensor systems which use an internal modulated light source to address the limitations of passive sensor systems have been developed (de Souza et al., 2010). Holland et al. (2012) described the radiometric principles of proximal active optical sensors, in particular, the inverse-square law of optics and how that influences operation and use of active sensors. Raun et al. (2002) first described the use of an active crop canopy sensor, measuring reflectance at 671 ± 6 and 780 ± 6 nm, to control the rate of fertilizer N applied to winter wheat. Sensor-based in-season N application improved nitrogen use efficiency (NUE) by >15% compared to traditional practices with uniform N rates. This system, developed at Oklahoma State University, became the GreenSeeker system (Trimble Inc., Sunnyvale CA, USA; Trimble, 2017a), which was one of the first commercially available active crop canopy sensors for N fertilizer management. The system relies on an in-field reference which is considered to be non-limiting for N in order to calibrate the sensor to specific cultivar and field conditions. The system produces NDVI values; in-field reference NDVI is divided by NDVI from the target area to generate a response index (RI - essentially the inverse of the SI described by Varvel et al., 2007). The RI is then used with an in-season estimate of yield (INSEY - Lukina et al., 2001) in a variety of crop- and locale-specific algorithms to generate an on-the-go fertilizer N rate. The GreenSeeker system can be used for N management of a variety of crops, but is mostly used for N management with corn at V6-V14 growth stages (Abendroth et al., 2011) and winter wheat at Feekes growth stages 4-6 (Feekes, 1941; Teal et al., 2006; Raun et al., 2002). Recently Trimble has developed a handheld version of this sensor for scouting purposes with an integrated power supply, a data logger and a GPS.

Another set of active crop canopy sensors is the Crop Circle suite of sensors (Holland Scientific, Lincoln, NE, USA). The initial sensor developed by Holland Scientific was the ACS-210, a two-band active sensor measuring reflectance at wavebands of 590 ± 5.5 and 880 ± 10 nm from an internal, modulated LED light source. Solari et al. (2010) described the development of an N rate recommendation algorithm for this sensor for corn at V11 and V15 growth stages, which was based on prior relationships of chlorophyll meter readings to N rate by Varvel et al. (2007). The ACS-210 is no longer commercially available and has been replaced by Holland Scientific with several sensors primarily for research purposes: the ACS-430, a three-band sensor (670, 730 and 780 nm); the ACS-470, a three-band sensor with interchangeable filters (user-selected between 420 and 800 nm) and the RapidScan CS-45, a handheld unit with the same optics as the ACS-430 and an integrated power supply, a data logger and a GPS.

The Holland Scientific set of sensors are intended primarily for research use. For commercial implementation, the technology is incorporated into the OptRx® crop sensor system (Ag Leader, Inc., Ames, IA, USA). The OptRx® system uses the same three-band sensor system as the Holland Scientific ACS-430, measuring reflectance at 670, 730 and 780-nm wavebands. One of the advantages of the OptRx® and other sensors using a red-edge band (730 nm) in place of the red band in the NDVI equation is the greater sensitivity of reflectance in the red-edge region to canopy chlorophyll and thus N content with high biomass plant canopies. Reflectance saturation occurs in the red region with high biomass crops such as corn as growth advances beyond LAI of ~ 2-3 (Thenkabail et al., 2000; Gitelson et al., 2005). The OptRx® system requires in-field calibration as does the GreenSeeker system and can be used with a managed high N reference area. However, the system is designed primarily for use with a universal N recommendation algorithm approach which can be adjusted by the user (Holland and Schepers, 2010) and using a virtual reference approach (Holland and Schepers, 2013). The virtual reference approach uses the top 5% of VI readings in an area of the field considered to represent the range of N status of the crop for the field and uses this value in the SI calculation. This approach precludes the need for the grower to establish a high N reference strip in the field.

Another active crop canopy sensor system available today is the Yara™ N-Sensor ALS (active light source), first offered commercially in 2005. A similar system is also marketed as the Topcon CropSpec™ crop canopy sensor in some countries. The N-Sensor ALS system uses a xenon flash lamp light source rather than an LED as in GreenSeeker and Crop Circle/OptRx systems. Other operating characteristics of the Yara™ N-Sensor ALS are similar to those of the passive sensor system N-Sensor™. The CropSpec™ system uses a laser light source, measuring reflectance in 730-740 and 800-810-nm bands, with other operating characteristics similar to those of the N-Sensor™ and N-Sensor™ ALS.

3 Current issues in sensor development

3.1 Crop canopy sensor limitations

The use of EM-based crop canopy sensors, particularly active sensors, has tremendous potential to improve the accuracy of N management through both refined rate and application timing. However, there are factors which can be sources of error and influence the accuracy of N fertilizer recommendations based on canopy sensor information. Samborski et al. (2009) provided an excellent review of potential sources of error for EM canopy sensors. These include sensor operating characteristics, such as sensor wavelengths and resulting VIs, sensor height and footprint/sensitivity issues; seasonal variation, related to algorithm calibration for specific growth stages; genotypic effects and most significantly, other stresses, such as water, nutrient, disease and insect stresses. The general assumption for use of proximal crop canopy sensors to manage N fertilization is that other sources of stress are absent or minimal or, even if present, will be equally present in reference areas as in the target areas, and thus, effects of other stresses will cancel out – assuming there are no interactions between N stress and other stresses. Water stress is perhaps of greatest concern, as water stress can be common, even in irrigated environments, and could confound sensor-based N management at critical times of the growing season. Researchers have investigated sensor N management when water and N stress are confounded (Shiratsuchi et al., 2011; Bronson et al., 2017). There is evidence that some VIs can correctly differentiate N stress in the presence of water stress; typically these are indices that include three wavebands in the equation: the Canopy Chlorophyll Content Index (CCCI, Barnes et al., 2000), the DATT index (Datt, 1999) and the Meris Terrestrial Chlorophyll Index (MTCI, Dash and Curran, 2004). None of these indices are currently available with commercial active crop canopy sensor systems.

3.2 Proximal sensing for stresses other than nitrogen

As is evident from the literature cited earlier, the majority of research and resulting commercial application of proximal crop sensing is for N fertilizer management for grain crops. There is substantial potential for adaptation of proximal crop sensing techniques for detecting other sources of stress, with several areas of ongoing research. However, there has been little commercial adoption of such sensor technologies to date. One technology which is commercially available for real-time detection and management of weeds is the WeedSeeker Spot Spray System (Trimble, 2017b). The WeedSeeker uses similar active canopy sensor technology to that found in the GreenSeeker canopy sensor, but adapted to detect weeds and then activate a herbicide application on-the-go. Sui et al. (2008) documented that the WeedSeeker system could accurately detect and manage weed infestations in cotton. Slaughter et al. (2007) provided a comprehensive review of autonomous weed detection and control research, and the use of various machine vision systems with associated weed recognition algorithms along with canopy reflectance for on-the-go weed control.

Disease and insect detection and intervention is another area in which proximal canopy sensing is being utilized. Franke and Menz (2007) have shown that early detection of disease onset in wheat can be accomplished with high-resolution multispectral imagery. Feng et al. (2016) found that reflectance in visible to red-edge bands (580–710 nm) could be used to monitor for powdery mildew infestation in mid-to-late growth stages of wheat. Hillnhütter et al. (2011) explored a range of VIs developed from hyperspectral imaging for their ability to detect the onset of damage to sugar beet from nematodes and fungal pathogens and noted that there was potential for field use of sensor technology to detect such stresses. Bravo et al. (2004) demonstrated that multispectral fluorescence images were useful for the detection of foliar disease in winter wheat. The primary proximal sensors commercially available today for disease detection are the suite of portable fluorometers from Force-A (Force-A, 2017). Sankaran and Ehsani (2012) found that the ratio of yellow fluorescence (UV excitation) to simple fluorescence (green excitation) was useful in the detection of Huanglongbing disease in citrus canopies.

Studies have used remote sensing technologies for detecting Russian wheat aphids in winter wheat (Mirik et al., 2012) and spider mites in corn (Fitzgerald et al., 2004). Liu et al. (2017) found that UV reflectance, used by birds along with visible wavelengths, could be useful in sensor detection of invertebrates on green leaves. Such sensor approaches have been little utilized with proximal sensors. Puig et al. (2015) did demonstrate that UAV-based RGB sensors could be used to detect white grub infestation in sorghum.

3.3 Sensor fusion and high-throughput phenotyping

An area of current research is the use of multiple proximal sensors to provide greater insight into plant characteristics than any single sensor – a concept termed 'sensor fusion'. Adamchuk et al. (2011) described the benefits of sensor fusion for precision agriculture, though their article primarily addressed sensor fusion for soil properties. They noted that the crop is an excellent bioindicator of variable growing conditions. They commented that '… a combination of conceptually different sensing techniques and integrating the subsequent data holds promise for providing more accurate property estimates, leading to more robust management and increased adoptability of sensor-based crop management'. A commercially available multi-sensor system for soil assessment is the Veris Technologies set of sensors which integrate soil apparent electrical conductivity (EC_a), soil pH and soil organic matter assessment in one platform (Veris Technologies, 2017). Data from multi-sensor systems can be integrated to generate variable-rate application maps for lime, fertilizer and seeding density. One of the earlier efforts to integrate multiple sensor information to assess crop properties was that of Shiratsuchi et al. (2009), who used a prototype system with a multispectral optical sensor for canopy reflectance, a thermal sensor for canopy temperature and an ultrasonic distance sensor for canopy height to assess crop N status. Vories et al. (2014) documented use of a similar system including multispectral canopy reflectance, an ultrasonic sensor for crop height and a thermal sensor for canopy temperature to assess water and N status of cotton. Weis et al. (2013) documented an effort to integrate four sensors [light detection and ranging (LIDAR), passive spectrometer, ultrasonic ranging and WeedSeeker™ active multispectral sensor] to regulate herbicide dosage in a study. Long and McCallum (2015) discussed the use of an on-combine set of sensors (mass flow yield monitor, in-line near-infrared spectrometer and LIDAR) to evaluate environmental stress in wheat at harvest. Sensor information documented crop yield, grain protein content and straw yield simultaneously. Lamb et al. (2014) used an airborne, high-intensity, LED-active multispectral sensor in conjunction with a passive thermal infrared sensor to map a cotton field at a height of 50 m above the canopy. They compared simple ratio (SR) VI maps from the active sensor to those developed from RapidEye satellite imagery (with the active sensor footprint and RapidEye pixel of similar dimensions), and found that the active sensor map closely resembled the RapidEye map. Combining information from the multispectral active and passive thermal sensors could accurately differentiate areas with full-to-partial plant cover from those with partial cover to bare areas.

One of the most active areas of investigation using sensor fusion is that of high-throughput phenotyping, primarily for accelerated identification of genetic traits for yield and stress tolerance for breeding purposes. This concept can involve applications that are lab- and greenhouse-oriented as well as field investigations. Busmeyer et al. (2013) described a multi-sensor platform for field phenotyping, which incorporated a colour camera, time-of-flight cameras, LIDAR, a hyperspectral imaging sensor and light curtain imaging systems. In this system, all sensors operated on a shrouded carrier to prevent exposure to ambient light. Consequently the height of plants over which the sensor could operate was limited; the authors documented repeatability of measurement results over triticale (*Triticosecale* Wittmack L.). The system could estimate values for plant moisture content, lodging, tiller density and biomass yield. Araus and Cairns (2014) provided an overview of high-throughput phenotyping platforms (HTTPs) and their capacity to inform genetic selection. In many cases, HTTPs utilize multispectral active crop canopy sensors, ultrasonic or LIDAR sensors for crop height and thermal sensors for canopy temperature. Wang et al. (2016) discussed the challenges with proper geolocation of data collected by HTTPs; with small plots normally used in breeding field trials, it is critical to accurately associate sensor data with precise location information. A relatively new integrated sensor designed specifically for phenotyping is the Crop Circle Phenom (Holland Scientific, 2017), which includes sensors for upwelling and downwelling photosynthetically active radiation, canopy temperature, air temperature, relative humidity, atmospheric pressure and canopy reflectance at wavelengths of 670, 730 and 780 nm.

4 Case studies

4.1 Case study 1: chlorophyll meter data - seasonal and management influences

This study illustrates the use of a simple crop canopy sensor, the chlorophyll meter, in understanding soil/fertilizer dynamics and resulting crop response during the growing season. This information is from a three-year study conducted on cooperating farmer's fields in Merrick County, Nebraska, USA. More detailed information from this study is available in Maharjan et al. (2016) and Peng et al. (2015). This particular data set is from the 2010 growing season. For clarity of illustration, only a subset of treatments is shown with this example. The cooperator planted corn on 15 April 2010, and nitrogen fertilizer was broadcast on 14 May 2010, shortly after crop emergence. To minimize the risk of ammonia volatilization in this study, fertilizers were lightly incorporated with a rotary tiller immediately after broadcast application. The study investigated several factors influencing crop yield and N fertilizer use efficiency, including N fertilizer source and rate. The most commonly used N fertilizer source in Nebraska is urea ammonium nitrate (UAN) solution – typically containing either 28% or 32% N. This fertilizer is favoured by farmers for a variety of reasons, including versatility in application methods and timing. At application, this fertilizer contains approximately 50% urea-N, 25% ammonium-N and 25% nitrate-N. Consequently, depending on the method of application and soil/ weather conditions, N from UAN solution may be susceptible to loss to ammonia volatilization, leaching or denitrification. The soil at the study site in 2010 was a loamy sand (mixed, superactive, mesic oxyaquic Haplustolls), and consequently the greatest risk for N loss was via nitrate leaching. The other fertilizer evaluated was a polymer-coated urea (PCU) product, ESN® (Environmentally Smart Nitrogen, Agrium, Inc., Calgary, Canada). The controlled-release of PCU provides gradual release of urea-N to the soil and crop with the intent of minimizing N loss through ammonia volatilization, nitrate leaching or denitrification and matching N supply to crop demand.

Figure 2 illustrates daily rainfall during the first month after fertilizer application. There were light precipitation events during the first two weeks after fertilization. A 17-mm rainfall event occurred on day 16, followed by 20 mm on day 20, 21 mm on day 25, 57 mm on day 27, 55 mm on day 29, intermixed with several other lighter rainfall events. By three weeks following fertilization, basically all urea in the UAN solution would have hydrolysed and nitrified to nitrate-N (Peng et al., 2015). Consequently, much of the N contained in UAN was highly susceptible to leaching on this soil with these conditions.

Chlorophyll (SPAD) meter readings were collected with a Minolta SPAD-502 meter in the middle of the uppermost fully expanded leaf until approximately the V14 growth stage (Abendroth et al. 2011), then from the ear leaf thereafter. Approximately 30 readings were collected from each plot, with mean values reported. Figure 3 illustrates chlorophyll meter data collected beginning at the V6 growth stage on 22 June 2010 – slightly more than two months after fertilizer application. It is at this growth stage or slightly thereafter that rapid uptake of N occurs with corn, between V8 and VT growth stages (Abendroth et al., 2011). Consequently, there were no significant differences in canopy chlorophyll status as influenced by N fertilizer source or rate at the V6 growth stage; only the unfertilized check was substantially lower in chlorophyll content at V6. This pattern was similar six days (28 June) later at V8. However, eight days later at V10, there began to be some separation in chlorophyll content among treatments – especially with UAN at the lower application rate of 168 kg N ha^{-1}. More separation among N sources and rates continued at subsequent growth stages, with chlorophyll content with PCU at either 168 or 280 kg N ha^{-1} substantially above that for UAN at either rate. These data indicate fairly early in

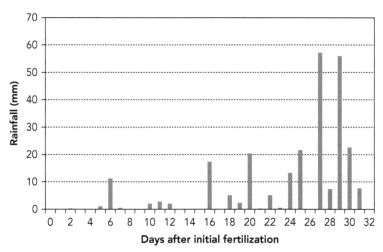

Figure 2 Daily precipitation for the first month after fertilization in 2010, Merrick County, Nebraska, USA.

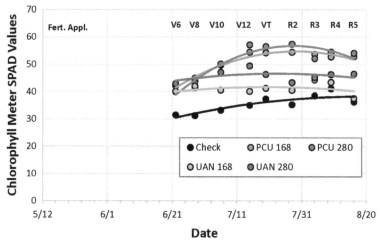

Figure 3 Chlorophyll meter readings as influenced by date/growth stage of corn during the 2010 growing season, fertilizer N source [urea ammonium nitrate (UAN) solution or polymer-coated urea (PCU)] and application rate (168 or 280 kg N ha⁻¹). Field location is in Merrick County, Nebraska, USA.

the growing season – at the V10 growth stage – that N availability to the crop was an issue with UAN. Chlorophyll levels with UAN tended to peak around the V12 growth stage, while chlorophyll levels with PCU continued to increase to VT and R2 growth stages. The gradual decline in canopy chlorophyll content during later reproductive growth stages is typical for corn, as nutrients increasingly are transferred to developing grain in the ear.

Figure 4 illustrates the impact of N fertilizer source and rate on grain yield in this study. At both 168 and 280 kg N ha⁻¹ rates, grain yield with PCU was approximately 4 Mg ha⁻¹ greater than for UAN at the same rates. These results confirm that substantial N loss from UAN reduced yield potential compared to PCU. Chlorophyll meter readings measured as early as the V10 growth stage indicated N deficiency with UAN treatments relative to PCU. Analyses by Peng et al. (2015) and Maharjan et al. (2016) indicate that hydrolysis of urea contained in UAN was mostly complete within a week after fertilization, followed by nitrification of fertilizer-derived ammonium to nitrate within the following couple of weeks. Consequently, it is most likely that rainfall during the period 16-30 days after fertilization leached much of the fertilizer N from UAN below the root zone on this coarse-textured site. Urea-N in PCU was gradually released through the growing season, and mostly protected against early season nitrate leaching.

This data set illustrates two attributes of proximal crop canopy sensing. For a research study, crop canopy sensing helps document interactions of treatments with soil and weather conditions over time. For a crop producer, proximal

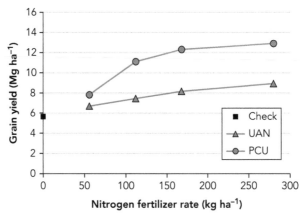

Figure 4 Grain yield as influenced by N source and rate, 2010, Merrick County, Nebraska, USA.

crop canopy sensing can provide invaluable management information during the growing season. If UAN fertilization at planting had been the standard management approach for a grower, canopy sensor information would have let the producer be informed of concerns about N supply at the V10 growth stage - early enough to intervene with supplemental fertilization to protect yield potential. One key issue here is that absolute chlorophyll meter readings are of little use - these can vary with hybrid and field conditions other than N supply. Consequently an in-field, non-limiting N reference, as suggested by Varvel et al. (2007), is necessary to accurately interpret chlorophyll meter information. A second key issue, as already noted, is that use of a handheld chlorophyll meter is impractical for management at the field scale, and other sensing approaches are advised.

4.2 Case study 2: use of an active crop canopy sensor for in-season N management and improved environmental stewardship

Corn producers in Nebraska, and generally in the United States, have steadily increased the efficiency with which they use N fertilizer. One parameter for measuring NUE which is relatively easily measured is partial factor productivity for N (PFP_N), which is the mass of grain produced per mass of fertilizer N applied. In Nebraska, PFP_N for corn has increased from around 35 kg grain kg N^{-1} in 1968 to around 65 kg grain kg N^{-1} in 2012 (Ferguson, 2015). This has been achieved through multiple factors, including improved genetics, changes in cropping systems, as well as more efficient fertilizer management practices. However, there has been little change in PFP_N since

2000, suggesting that current fertilizer management practices may be limiting further increases in efficiency, and that new approaches to N management are needed to continue increasing NUE. While active crop canopy sensors have been on the market in the United States for over a decade, there has been limited adoption of this technology for N fertilizer management to date. As with any new technology, there can be a range of factors influencing producer decisions to adopt, including cost of the technology, impacts on profit, unfamiliarity with the technology or uncertainty about its application to specific production systems. In order to address some of these issues relative to active canopy sensor N management for corn, a project was started in 2015 in Nebraska, USA, as part of an on-farm research initiative. The project, called Project SENSE (Sensors for Efficient Nitrogen Use and Stewardship of the Environment), is a collaborative effort of the University of Nebraska-Lincoln, the Nebraska corn board and five natural resource districts (NRDs) in Nebraska (NRDs are governmental agencies with boundaries organized along watersheds charged with management of soil and water resources in their district). The objective of Project SENSE is to encourage improved NUE through in-season fertilization. One of the tools promoted by the project is the use of active crop canopy sensor-based in-season application. The project gives cooperating producers the opportunity to see an active canopy sensor-equipped high clearance applicator operate on their fields and to compare their standard N management approaches to sensor-based N management in a scientific manner. Project SENSE is a component of the Nebraska On-Farm Research Network, an organization of growers who conduct scientifically valid

Figure 5 Project SENSE high clearance applicator (Hagie DTS-10) equipped with OptRx® active crop canopy sensors. Typically an applicator of this width would use 2–3 OptRx sensors; this applicator is equipped with 7 sensors, to allow flexibility with treatment strip width and to collect supplemental reflectance data for research purposes.

field-scale research on their fields and share results in annual conferences and on the project website (Nebraska On-Farm Research Network, 2017).

Approximately 20 producers participate in Project SENSE each year. A typical site compares the grower's standard N management to sensor-based N management in field - length strips, usually with six replications of each treatment. The widths of treatment strips are based on the cooperator's fertilizer application and harvest equipment and typically are 8–16 rows (6–12 m) wide. Growers use their choice of N fertilizer application rate and timing; most use split application timing, with some fertilizer applied at or prior to planting and the remainder side-dressed early in the growing season. Practically all cooperators use uniform N application rates. A base rate of 70–80 kg N ha^{-1} was applied to sensor-based treatment strips by the cooperator at or prior to planting. For Nebraska soil and weather conditions, this amount of fertilizer N, coupled with soil-supplied N, allows the crop to reach the V8–V9 growth stage without yield-limiting N stress. The project uses the AgLeader OptRx® active crop canopy sensor system, including the standard N recommendation algorithm approach included with the sensors. This includes a virtual reference approach (Holland and Schepers, 2013). To determine the reference, an area considered to represent the range of crop N status for the field is driven over by the high clearance applicator (Fig. 5) and sensed to build a virtual reference data set. Once the virtual reference is established, sensor-based treatments were applied. In-season application occurred between V8 and V14 growth stages, with a target of application as close as possible to V9 growth stage. Otherwise management was at the grower's discretion, including fertilizers other than N, hybrid and seed density, use of pesticides and irrigation. The cooperator harvested each treatment strip with a yield-mapping combine. Results were then analysed statistically to determine treatment effects on fertilizer use, grain yield, NUE and partial profit.

Figure 6a–b illustrates results from one 2016 site. Figure 6a is a map of Normalized Difference Red Edge Vegetation Index (NDRE, Barnes et al., 2000) reflectance values for sensor-based treatment strips; the grower treatment strips are randomized among the sensor-based treatment strips. This illustrates a range in crop N status across the field, primarily influenced by soil variation. Figure 6b shows the resulting side-dress N rate calculation based on canopy reflectance relative to reference values within the study area; side-dress N application rates ranged from 34 kg N ha^{-1} (the minimum rate applied) to 204 kg N ha^{-1}. Figure 6c shows resulting grain yield from all treatment strips, including the cooperator's uniform rate treatment strips. Figure 6d shows the mean results comparing the grower treatment to sensor-based treatment. The average fertilizer N rate for the sensor-based treatment was 40 kg N ha^{-1} less than the grower rate, while grain yield was statistically the same, at around 9.4 Mg ha^{-1}. In this case, profit increased with sensor-based treatment due

(a)

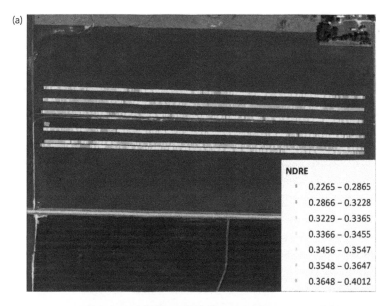

NDRE
- 0.2265 – 0.2865
- 0.2866 – 0.3228
- 0.3229 – 0.3365
- 0.3366 – 0.3455
- 0.3456 – 0.3547
- 0.3548 – 0.3647
- 0.3648 – 0.4012

(b)

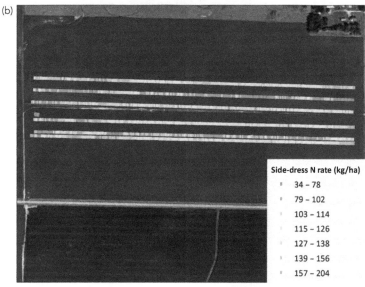

Side-dress N rate (kg/ha)
- 34 – 78
- 79 – 102
- 103 – 114
- 115 – 126
- 127 – 138
- 139 – 156
- 157 – 204

Figure 6 (a) Normalized Difference Red Edge (NDRE) for Project SENSE site 16BK, 2016. NDRE is shown only for crop canopy sensor-based treatment strips. (b) Side-dress fertilizer N rate for Project SENSE site 16BK, 2016. Nitrogen rate is shown only for crop canopy sensor-based treatment strips. (c) Grain yield for Project SENSE site 16BK, 2016. Yield is shown for both sensor-based treatment strips and grower treatment strips. (d) Total mean fertilizer N applied (red markers) and mean grain yield (bars) for the grower and sensor-based treatment strips, Project SENSE site 16BK, 2016.

(c)

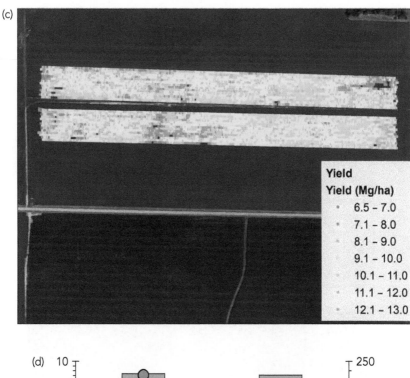

Yield
Yield (Mg/ha)

- 6.5 – 7.0
- 7.1 – 8.0
- 8.1 – 9.0
- 9.1 – 10.0
- 10.1 – 11.0
- 11.1 – 12.0
- 12.1 – 13.0

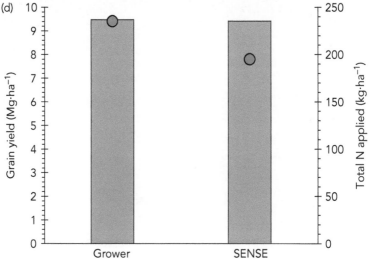

Figure 6 (*Continued*)

(a)

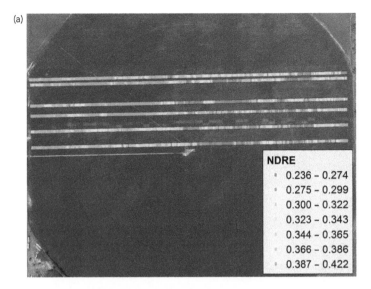

NDRE
- 0.236 – 0.274
- 0.275 – 0.299
- 0.300 – 0.322
- 0.323 – 0.343
- 0.344 – 0.365
- 0.366 – 0.386
- 0.387 – 0.422

(b)

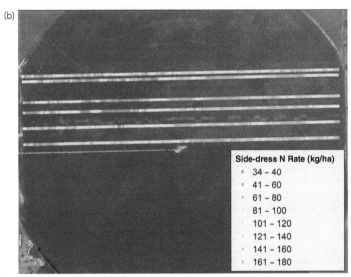

Side-dress N Rate (kg/ha)
- 34 – 40
- 41 – 60
- 61 – 80
- 81 – 100
- 101 – 120
- 121 – 140
- 141 – 160
- 161 – 180

Figure 7 (a) Normalized Difference Red Edge (NDRE) for Project SENSE site 16KF, 2016. NDRE is shown only for crop canopy sensor-based treatment strips. (b) Side-dress fertilizer N rate for Project SENSE site 16KF, 2016. Nitrogen rate is shown only for crop canopy sensor-based treatment strips. (c) Grain yield for Project SENSE site 16KF, 2016. Yield is shown for both sensor-based treatment strips and grower treatment strips. (d) Total mean fertilizer N applied (red markers) and mean grain yield (bars) for the grower and sensor-based treatment strips, Project SENSE site 16KF, 2016.

(c)

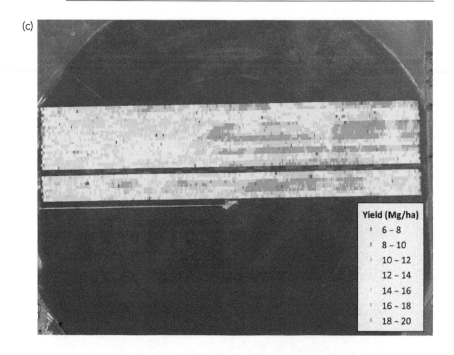

Yield (Mg/ha)
- 6 – 8
- 8 – 10
- 10 – 12
- 12 – 14
- 14 – 16
- 16 – 18
- 18 – 20

(d)

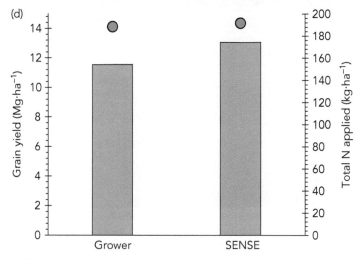

Figure 7 (*Continued*)

Table 1 Mean fertilizer N application, grain yield, NUE and profit, Project SENSE sites (n = 17 for 2015, n = 20 for 2016). Marginal net return based on: 2015 N price of $1.44 kg N^{-1} and corn price of $0.14 kg^{-1}; 2016 N price of $1.00 kg^{-1} and corn price of $0.12 kg^{-1}

	2015		2016	
	Grower treatment	Sensor-based treatment	Grower treatment	Sensor-based treatment
Total N applied (kg ha^{-1})	218	194	212	174
Grain yield (Mg ha^{-1})	14.25	13.94	12.68	12.49
PFP$_N$ (kg grain kg N^{-1})	65	80	63	75
Marginal net return ($ ha^{-1})	1733.44	1752.59	1310.19	1327.58

to the use of less fertilizer N, while yield remained the same. Figure 7a–d illustrates the same information (NDRE, side-dress N application rate, grain yield and treatment mean total N applied and grain yield) for a second site in 2016. At this site, total overall N applied was basically the same between the grower and sensor-based treatments – around 170 kg N ha^{-1}. However, primarily due to more timely application, the sensor-based treatments produced on average 2.5 Mg ha^{-1} greater yield than the grower treatment.

Overall results from the project for 2015 and 2016 are shown in Table 1. In both years, sensor-based N application significantly reduced the total amount of N applied compared to grower practices by 44 kg N ha^{-1} in 2015 and 38 kg N ha^{-1} in 2016. This resulted in slightly reduced average grain yields compared to grower standard practices in both years, by 0.31 Mg ha^{-1} in 2015 and 0.19 Mg ha^{-1} in 2016. However, fertilizer NUE (PFP$_N$) increased substantially with sensor-based application. Grower treatments produced an average of 65 kg grain kg N^{-1} in 2015 and 63 kg grain kg N^{-1} in 2016. Sensor-based PFP$_N$ was 80 kg grain kg N^{-1} in 2015 and 75 kg grain kg N^{-1} in 2016. Perhaps most importantly to the growers, marginal net return was increased both years with sensor-based management – by $19.14 ha^{-1} in 2015 and $17.39 ha^{-1} in 2016. Depending on the type of sensor and applicator control system a grower would purchase, it would typically require use over 218–437 ha over four years to pay for the sensor system purchase at current equipment prices of $15 000–$30 000, and commodity prices of $0.12 kg^{-1} corn and $1.00 kg^{-1} N.

While overall increased profit is of concern to the farmer, reduced environmental impacts from fertilizer use are of concern to society. Grower sites in Project SENSE are located in areas of Nebraska with elevated groundwater nitrate-N levels, resulting from past overuse of N fertilizers coupled with

Figure 8 Manual operation of a UAS with RapidScan CS-45 active sensor.

excessive irrigation resulting in nitrate leaching. Typically, groundwater nitrate-N concentrations in these areas can range from 10 to 30 mg L^{-1} or more (irrigation water nitrate-N credits were accounted for in calculating sensor-based N applications at these sites). The reduction of 30–40 kg N ha^{-1} or more at these sites, with little impact on grain yield and an increase in profit, with the use of sensor-based N application will have a positive impact on groundwater quality over time.

4.3 Case study 3: use of an unmanned, aerial system-mounted (UAS-mounted) active crop canopy sensor

As previously mentioned, the rapid growth in availability of UAS with associated airframes and sensor packages is blurring the traditional boundary between proximal and remote sensing. UAS can operate in close proximity to the crop canopy, allowing spatial and temporal resolution well beyond that achievable with remote sensing systems based on aircraft or satellites. Optical sensing systems used with UAS are typically passive sensors, which restrict their utility based on time of day and weather conditions. There are several advantages to the use of active sensors with a UAS; they can be used regardless of sun conditions, day or night; they are not imaging sensors, so images do not need to be processed, geo-rectified or stitched; data are thus immediately available for use, with geographic position information, without extensive processing. However, light sources used with active canopy sensors are relatively of low power, and sensors may have limited sensitivity beyond a few metres above

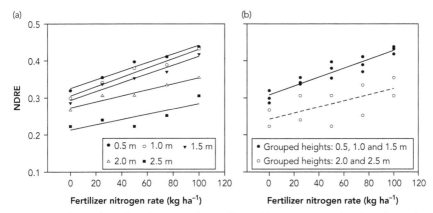

Figure 9 Relationship of UAS-mounted RapidScan CS-45 at various heights above the crop canopy (0.5, 1.0, 1.5, 2.0 and 2.5 m) to fertilizer N rate and NDRE (a) and grouped heights to fertilizer N rate and NDRE (b).

the crop canopy. GreenSeeker and OptRx sensors, for example, typically are designed for use within a couple of metres or less from the crop canopy. Previous examples of an airborne active sensor are provided by Lamb et al. (2009) who used a standard CropCircle™ active sensor mounted to an aircraft and flown at a height of 3–5 m above a sorghum canopy, and Lamb et al. (2011) who used an active sensor with a high-powered LED system (Raptor™, Holland Scientific) to map NDVI at heights of 15–45 m above a wheat canopy. In both studies, data collected with airborne active sensors showed good correlation with data collected from a passive multispectral aerial sensor in the first study, or with a ground-based active sensor in the second study.

A project conducted in Nebraska beginning in 2014 evaluated the potential for use of active canopy sensors with UAS. The project used a UAS-mounted Holland Scientific RapidScan CS-45 (Fig. 8), modified such that it would log reflectance information and position every second, rather than averaging values over a period of time. The RapidScan CS-45 sensor is intended for handheld use at a distance of 0.3–3 m above the crop canopy. Krienke et al. (2017) evaluated the sensitivity of the UAS-mounted RapidScan to height above the crop canopy and investigated the relationships between fertilizer N rate, NDRE and corn grain yield. The study found that up to a height of 1.5 m above a turfgrass canopy (selected due to its lack of complexity and uniform depth of canopy relative to a maize canopy), sensor readings were relatively similar. However, above 1.5 m, the slope of NDRE relative to fertilizer N rate was different from that below 1.5 m (Fig. 9). For this particular sensor they concluded that use at heights of 1.5 m or less above the canopy would produce reliable estimates of crop N status. In a separate trial, they evaluated the relationship of active sensor NDRE above a corn study with varying fertilizer N rates. The study had

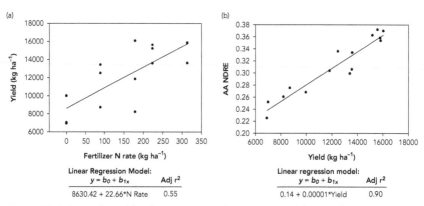

Figure 10 Relationship of fertilizer N rate and corn grain yield (a) and grain yield and NDRE (b).

significant scatter in the relationship between fertilizer N rate and grain yield (Fig. 10a), due to small-scale variation in soil N supply even within small plots. However, the relationship between NDRE from the UAS-mounted active sensor at the V11 growth stage and final grain yield was quite strong (Fig. 10b).

This study illustrates the potential for active sensor use with a UAS. The sensor used in the above study was not designed with a light source intended for UAS operation. Maintaining a constant height of 1.0–1.5 m as in this study manually is not practical with current elevation control systems available for UAS. However, a more intense light source would allow greater distance between the UAS sensor and the crop canopy while still obtaining accurate canopy reflectance information. Also, ranging sensors, either ultrasonic or LIDAR, could be incorporated into the flight control system of the UAS to maintain an accurate height above the crop canopy. Such a UAS system could operate at any time of the day – or night, if regulations allowed, at a height of a few metres above the canopy in airspace with little concern for air traffic, other than perhaps agricultural spray planes. The extended times of data collection could also allow a UAS equipped with an active sensor to operate at dawn or dusk, when wind speeds in many areas are typically lower, allowing for more stable UAS operation and improved data accuracy.

5 Conclusion: sustainability and environmental implications

The concept of proximal crop sensing is directly related to improved sustainability and reduced environmental impact of producing crops. One of the primary concepts of precision agriculture is site-specific management – adjusting crop inputs, or even the crop cultivar or species, based on the capacity of the underlying soil resource. Both proximal and remote crop sensing enable such spatial management, but proximal sensing adds the capacity to better manage temporal variation as well, due to inherent temporal constraints with remote sensing via aircraft or satellite platforms. Both remote and proximal sensors allow growers to monitor the crop during the growing season and adjust management according to such information. The potential for more judicious use of resources such as soil and water, and crop inputs such as fertilizer, pesticides or tillage seems evident when growers can detect and manage emerging issues in a very timely and site-specific manner, rather than use of blanket, prophylactic amendments which may not be necessary and have the potential for environmental degradation when used in excess.

6 Future trends for research

As indicated earlier, most proximal crop sensors today are either passive sensors using optical reflectance from the canopy using the sun as a light source or active sensors using internal, modulated light sources. There appears to be significant potential for research and development of active canopy sensors for use with UAS, for rapid, more timely spatial surveys of crops and to expand properties of interest beyond N, to water, other nutrient stresses, as well as disease and insect stress. There is significant potential to further explore multispectral fluorescence as an indicator of specific stresses. The mid-infrared and thermal EM regions have not yet been widely studied as indicators of stress, other than for water stress, primarily due to the lack of lower cost sensors in these wavelength ranges. Finally, there seems to be significant potential for sensors other than EM sensors, for example, pheromone or spore detectors for the onset of insect or disease or other chemo-detectors in general. These seem especially suited to UAS use, if detection of a stress indicator can be easily coupled with geographic position within a field, allowing rapid mapping of field for stress indicators.

7 Where to look for further information

Further information on proximal crop sensing can be found through conferences sponsored by the International Society of Precision Agriculture (ISPA; www. ispa.org) and from research journals, such as Precision Agriculture (springer. com/journal/11119).

8 References

Abendroth, L. J., R. W. Elmore, M. J. Boyer and S. K. Marlay (2011), 'Corn growth and development', PMR 1009. Ames, IA, USA, Iowa State University.

Adamchuk, V., R. Viscarra Rossel, K. Sudduth and P. Lammers (2011), 'Sensor fusion for precision agriculture', In: C. Thomas (Ed.), *Sensor Fusion - Foundation and Applications*. InTech Inc., Vienna, Austria. DOI: 10:5772/680. https://www.intechopen.com/books/sensor-fusion-foundation-and-applications/.

Ag Leader (2017), OptRx® Crop Sensors. http://www.agleader.com/products/directcommand/optrx-crop-sensors/ (Accessed 20 April 2017).

Araus, J. L. and J. E. Cairns (2014), 'Field high-throughput phenotyping: The new crop breeding frontier', *Trends in Plant Science* 19: 52-61. http://dx.doi.org/10.1016/j.tplants.2013.09.008.

Barnes, E. M., T. R. Clarke, S. E. Richards, P. D., Colaizzi, J. Haberland and M. Kostrzewski (2000), 'Coincident detection of crop water stress, nitrogen status and canopy density using ground-based multi-spectral data [CD-ROM]', In: P. C. Robert, R. H Rust and W. E. Larson (Eds), *Proceedings 5th International Conference on Precision Agriculture*. Madison, WI, USA, ASA, CSSA, and SSSA.

Berntsen, J., A. Thomas, K. Schelde, O. M. Hanse, L. Knudsen, N. Broge, H. Hougaard and R. Hørfarter (2006), 'Algorithms for sensor-based redistribution of nitrogen fertilizer in winter wheat', *Precision Agriculture* 7: 65-83.

Blackmer, T. M. and J. S. Schepers (1995), 'Use of a chlorophyll meter to monitor nitrogen status and schedule fertigation for corn', *Journal of Production Agriculture* 8: 56-60.

Bravo, C., D. Moshou, R. Oberti, J. West and A. McCartney, L. Bodria and H. Ramon (2004), 'Foliar disease detection in the field using optical sensor fusion', *Agricultural Engineering International: The CIGR Journal of Scientific Research and Development* VI: 1-13, manuscript FP 04 008.

Bréda, N. J. J. (2003), 'Ground-based measurements of leaf area index: A review of methods, instruments and current controversies', *Journal of Experimental Botany* 54: 2403-17.

Bronson, K. F., J. W. White, M. M. Conley, D. J. Hunsaker, K. R. Thorp, A. N. French, B. E. Mackey and K. H. Holland (2017), 'Active optical sensors in irrigated durum wheat: Nitrogen and water effects', *Agronomy Journal* 109: 1-12.

Bundy, L. G. and T. W. Andraski (2004), 'Diagnostic tests for site-specific nitrogen recommendations for winter wheat', *Agronomy Journal* 96: 608-14.

Busemeyer, L., D. Mentrup, K. Möller, E. Wunder, K. Alheit, V. Hahn, H. P. Maurer, J. C. Reif, T. Würschum, J. Müller, F. Rahe and A. Ruckelshausen (2013), 'BreedVision - a multi-sensor platform for non-destructive field-based phenotyping in plant breeding', *Sensors* 13: 2830-47. doi:10.3390/s130302830.

Dash, J. and P. J. Curran (2004), 'The MERIS terrestrial chlorophyll index', *International Journal of Remote Sensing* 25: 5403-13. doi:10.1080/0143116042000274015.

Datt, B. (1999), 'A new reflectance index for remote sensing of chlorophyll content in higher plants: Tests using Eucalyptus leaves', *Journal of Plant Physiology* 154: 30-6. doi:10.1016/S0176-1617(99)80314-9.

De Souza, E. G., P. C. Scharf and K. A. Sudduth (2010), 'Sun position and cloud effects on reflectance and vegetation indices of corn', *Agronomy Journal* 102: 734-44.

Dworak, V., J. Selbeck and D. Ehlert (2011), 'Ranging sensors for vehicle-based measurement of crop stand and orchard parameters: A review', *Transactions of the American Society of Agricultural and Biological Engineers* 54(4): 1497-510.

Ehlert, D. and K. H. Dammer (2006), 'Widescale testing of the Crop-meter for site-specific farming', *Precision Agriculture* 7: 101-15.

Feekes, W. (1941), 'De tarwe en haar mileu [Wheat and its environment]', *Verslagen van de Technische Tarwe Commissie* (English summary) 17: 523–888.

Feng, W., W. Shen, L. He, J. Duan, B. Guo, Y. Li, C. Wang and T. Guo (2016), 'Improved remote sensing detection of wheat powdery mildew using dual-green vegetation indices', *Precision Agriculture* 17: 608–827. doi: 10.1007/s11119-016-9440-2.

Ferguson, R. B. (2015), 'Groundwater quality and nitrogen use efficiency in Nebraska's Central Platte River Valley', *Journal of Environmental Quality* 44: 449–59.

Fitzgerald, G., S. Maas and W. Detar (2004), 'Spider mite detection and canopy component mapping in cotton using hyperspectral imagery and spectral mixture analysis', *Precision Agriculture* 5: 275–89.

Follett, R. H., R. F. Follett and A. D. Halvorson (1992), 'Use of a chlorophyll meter to evaluate the nitrogen status of dryland winter wheat', *Communications in Soil and Plant Analysis* 23: 687–97.

Force-A (2017), Dualex Scientific™, Multiplex Research™, Multiplex 330™. http://www.force-a.com/en/ (Accessed 17 April 2017).

Franke, J. and G. Menz (2007), 'Multi-temporal wheat disease detection by multi-spectral remote sensing', *Precision Agriculture* 8: 161–72.

Gitelson, A. A., A. Viña, V. Ciganda, D. C. Rundquist and T. J. Arkebauer (2005), 'Remote estimation of canopy chlorophyll content in crops', *Geophysical Research Letters* 32, L08403, doi:10.1029/2005GL022688.

Hillnhütter, C., A. Mahlein, R. Sikora and E. Oerke (2011), 'Use of imaging spectroscopy to discriminate symptoms caused by *Heterodera schachtii* and *Rhizoctonia solani* on sugar beet', *Precision Agriculture* 13: 17–32.

Holland, K. H. and J. S. Schepers (2010), 'Derivation of a variable rate nitrogen application model for in-season fertilization of corn', *Agronomy Journal* 102: 1415–24.

Holland, K. H., D. W. Lamb and J. S. Schepers (2012), 'Radiometry of proximal active optical sensors (AOS) for agricultural sensing', *IEEE Journal of Selected Topics in Applied Earth Observations and Remote Sensing* 5: 1793–801.

Holland, K. H. and J. S. Schepers (2013), 'Use of a virtual-reference concept to interpret active crop canopy sensor data', *Precision Agriculture* 14: 71–85.

Holland Scientific (2017), Crop Circle Phenom. http://hollandscientific.com/portfolio/crop-circle-phenom/ (Accessed 8 May 2017).

Jørgensen, J. R. and R. N. Jørgensen (2007), 'Uniformity of wheat yield and quality using sensor assisted application of nitrogen', *Precision Agriculture* 8: 63–73.

Konica Minolta (2017), Chlorophyll meter SPAD-502Plus. https://www.konicaminolta.com/instruments/download/catalog/color/pdf/spad502plus_catalog_eng.pdf (Accessed 14 April 2017).

Krienke, B., R. Ferguson and B. Maharjan (2017), 'Using an unmanned aerial vehicle to evaluate nitrogen variability and height effect with an active crop canopy sensor', *Precision Agriculture* 18:900-15.

Lamb, D., M. Trotter and D. Schneider (2009), 'Ultra low-level airborne (ULLA) sensing of crop canopy reflectance; A case study using a CropCircle™ sensor', *Computers and Electronics in Agriculture* 69: 86–91.

Lamb, D., D. Schneider, M. Trotter, M. Schafer and I. Yule (2011), 'Extended-altitude aerial mapping of crop NDVI using an active optical sensor; A case study using a Raptor™ sensor over wheat', *Computers and Electronics in Agriculture* 77: 69–73.

Lamb, D., D. Schneider and J. Stanley (2014), 'Combination active optical and passive thermal infrared sensor for low-level airborne crop sensing', *Precision Agriculture* 15: 523-31.

Lee, K. H. and R. Ehsani. (2009), 'A laser scanner based measurement system for quantification of citrus tree geometric characteristics', *Applied Engineering in Agriculture* 25(5): 777-88.

Liu, H., S. Lee and J. Chahl (2017), 'An evaluation of the contribution of ultraviolet in fused multispectral images for invertebrate detection on green leaves', *Precision Agriculture* 18: 667-83.

Long, D. and J. McCallum (2015), 'On-combine, multi-sensor data collection for post-harvest assessment of environmental stress in wheat', *Precision Agriculture* 16: 492-504.

Lukina, E. V., K. W. Freeman, K. J. Wynn, W. E. Thomason, R. W. Mullen, M. L. Stone, J. B. Solie and W. R. Raun (2001), 'Nitrogen fertilization optimization algorithm based on in-season estimates of yield and plant nitrogen uptake', *Journal of Plant Nutrition* 24: 885-98.

Maharjan, B., R. B. Ferguson and G. P. Slater (2016), 'Polymer-coated urea improved corn response compared to urea-ammonium nitrate when applied on a coarse-textured soil', *Agronomy Journal* 108: 509-18.

Markwell, J., J. C. Osterman and J. L. Mitchell (1995), 'Calibration of the Minolta SPAD-502 leaf chlorophyll meter', *Photosynthesis Research* 46: 467-72.

Mirik, M., R. Ansley, G. Michels Jr. and N. Elliott (2012), 'Spectral vegetation indices selected for quantifying Russian wheat aphid (*Diuraphis noxia*) feeding damage in wheat (*Triticum aestivum* L.)', *Precision Agriculture* 13: 501-16.

Nebraska On-Farm Research Network (2017), http://cropwatch.unl.edu/on-farm-research (Accessed 30 April 2017).

Ortuzar-Iragorri, M. A., A. Alonso, A. Castellón, G. Besga, J. M. Estavillo and A. Aizpurua (2005), 'N-Tester use in soft winter wheat: Evaluation of nitrogen status and grain yield prediction', *Agronomy Journal* 97: 1380-9.

Peng, X., B. Maharjan, C. Yu, A. Su, V. Jin and R. B. Ferguson (2015), 'A laboratory evaluation of ammonia volatilization and nitrate leaching following nitrogen fertilizer application on a coarse-textured soil', *Agronomy Journal* 107: 871-9.

Piekielek, W. P. and R. H. Fox (1992), 'Use of a chlorophyll meter to predict sidedress nitrogen requirements for maize', *Agronomy Journal* 84: 59-65.

Puig, E., F. Gonzalez, G. Hamilton and P. Grundy (2015), 'Assessment of crop insect damage using unmanned aerial systems: A machine learning approach', In: T. Weber, M. J. McPhee and R. S. Anderssen (Eds), *MODSIM2015, 21st International Congress on Modelling and Simulation*. Modelling and Simulation Society of Australia and New Zealand, Gold Coast, Australia, December 2015, pp. 1420-6. ISBN: 978-0-9872143-5-5. http://www.mssanz.org.au/modsim2015/F12/puig.pdf (Accessed 3 May 2017).

Raun, W. R., J. B. Solie, G. V. Johnson, M. L. Stone, E. V. Lukina, W. E. Thomason and J. S. Schepers (2001), 'In-season prediction of potential grain yield in winter wheat using canopy reflectance', *Agronomy Journal* 93: 131-8.

Raun, W. R., J. B. Solie, G. V. Johnson, M. L. Stone, R. W. Mullen, K. W. Freeman, W. E. Thomason and E. V. Lukina (2002), 'Improving nitrogen use efficiency in cereal grain production with optical sensing and variable rate application', *Agronomy Journal* 94: 815-20.

Rouse Jr., J. W., R. H. Haas, J. A. Schell and D. W. Deering (1974), 'Monitoring vegetation systems in the Great Plains with ERTS', In: *Third Earth Resources Technology Satellite-1 Symposium- Volume I: Technical Presentations. NASA SP-351*, Washington, DC, 10-24 December 1973. Washington, DC, NASA, pp. 309-17.

Samborski, S. M., N. Tremblay and E. Fallon (2009), 'Strategies to make use of plant sensor-based diagnostic information for nitrogen recommendations', *Agronomy Journal* 101: 800-16.

Sankaran, S. and R. Ehsani (2012), 'Detection of Huanglongbing Disease in citrus using fluorescence spectroscopy', *Transactions of the American Society of Agricultural and Biological Engineers* 55(1): 313-20.

Scharf, P. C., S. M. Brouder and R. G. Hoeft (2006), 'Chlorophyll meter readings can predict nitrogen need and yield response of corn in the North-Central USA', *Agronomy Journal* 98: 655-85.

Shapiro, C. A., D. D. Francis, R. B. Ferguson, G. W. Hergert, T. M. Shaver and C. S. Wortmann (2013), 'Using a chlorophyll meter to improve N management', NebGuide G1632. University of Nebraska-Lincoln, Lincoln, Nebraska, USA. http://extensionpublications. unl.edu/assets/pdf/g1632.pdf (Accessed 14 April 2017).

Shiratsuchi, L. S., R. B. Ferguson, V. I. Adamchuk, J. F. Shanahan and G. P. Slater (2009), 'Integration of ultrasonic and active canopy sensors to estimate the in-season nitrogen content for corn', In: *Proceedings of the 39th North Central Extension-Industry Soil Fertility Conference*. Norcross Georgia, USA, International Plant Nutrition Institute, pp. 182-8.

Shiratsuchi, L. S. (2011), 'Integration of plant-based canopy sensors for site-specific nitrogen management', *Theses, Dissertations and Student Research in Agronomy and Horticulture*, University of Nebraska-Lincoln Digital Commons. http:// digitalcommons.unl.edu/agronhortdiss/36 (Accessed 14 April 2017).

Shiratsuchi, L., R. Ferguson, J. Shanahan, V. Adamchuk, D. Rundquist, D. Marx and G. Slater (2011), 'Water and nitrogen effects on active canopy sensor vegetation indices', *Agronomy Journal* 103: 1815-26.

Slaughter, D. C., D. K. Giles and D. Downey (2007), 'Autonomous robotic weed control systems: A review', *Computers and Electronics in Agriculture* 61: 63-78.

Solari, F., J. F. Shanahan, R. B. Ferguson and V. I. Adamchuk (2010), 'An active sensor algorithm for corn nitrogen recommendations based on a chlorophyll meter algorithm', *Agronomy Journal* 102: 1090-8.

Steinberg, S., C. H.M. Van Bavel, and M. J. McFarland (1989), 'A gauge to measure mass flow rate of sap in stem and trunks of woody plants', *Journal of the American Society of Horticultural Science* 114: 466-72.

Sui, R., J. A. Thomasson, J. Hanks and J. Wooten (2008), 'Ground-based sensing system for weed mapping in cotton', *Computers and Electronics in Agriculture* 60: 31-8.

Teal, R. K., B. Tubana, K. Girma, K. W. Freeman, D. B. Arnall, O. Walsh and W. R. Raun (2006), 'In-season prediction of corn grain yield potential using normalized difference vegetation index', *Agronomy Journal* 98: 1488-94.

Thenkabail, P. S., R. B. Smith and E. De Pauw (2000), 'Hyperspectral vegetation indices and their relationships with agricultural crop characteristics', *Remote Sensing of the Environment* 71: 158-82.

Tremblay, N., Z. Wang and C. Bélec (2007), 'Evaluation of the Dualex for the assessment of corn nitrogen status', *Journal of Plant Nutrition* 30(9): 1355-69.

Tremblay, N., Z. Wang, B. Ma, C. Belec and P. Vigneault (2009), 'A comparison of crop data measured by two commercial sensors for variable-rate nitrogen application', *Precision Agriculture* 10: 145–61.

Trimble (2017a), https://agriculture.trimble.com/precision-ag/products/greenseeker/ (Accessed 19 April 2017).

Trimble (2017b), https://agriculture.trimble.com/precision-ag/products/weedseeker/ (Accessed 29 April 2017).

Varvel, G. E., W. W. Wilhelm, J. F. Shanahan and J. S. Schepers (2007), 'An algorithm for corn nitrogen recommendations using a chlorophyll meter based sufficiency index', *Agronomy Journal* 99: 701–6.

Veris Technologies (2017), http://www.veristech.com/ (Accessed 5 May 2017).

Vories, E., A. Jones, J. Sudduth, S. Drummond and N. Benson (2014), 'Sensing nitrogen requirements for irrigated and rainfed cotton', *Applied Engineering in Agriculture* 30: 707–16.

Wallihan, E. F. (1973), 'Portable reflectance meter for estimating chlorophyll concentration in leaves', *Agronomy Journal* 65: 659–62.

Wang, X., K. Thorp, J. White, A. French and J. Poland (2016), 'Approaches for geospatial processing of field-based high-throughput plant phenomics data from ground vehicle platforms', *Transactions of the American Society of Agricultural and Biological Engineers* 59: 1053–67.

Weis, M., D. Andujar, G. G. Peteinatos and R. Gerhards (2013), 'Improving the determination of plant characteristics by fusion of four different sensors', In: J. V. Stafford (Ed.), *Precision Agriculture'13, Proceedings of the 9th European Conference on Precision Agriculture*. The Netherlands, Wageningen Academic Publishers, pp. 63–9.

Yara International ASA (2017), http://yara.com/ (Accessed 17 April 2017).

Zebarth, B. J., H. Rees, N. Tremblay, P. Fournier and B. Leblon (2003), 'Mapping spatial variation in potato nitrogen status using the "N-Sensor"', *Acta Horticulturae* 627: 267–73.

Chapter 2

Advances in proximal sensors to detect crop health status in horticultural crops

Catello Pane, CREA – Research Centre for Vegetable and Ornamental Crops, Italy

1 Introduction

Digital technologies can assist farmers in managing plant diseases of horticultural crops in various ways. They can increase the ability to monitor the incidence of disease within both extensive and intensive types of cultivation, provide important information on the spatial/temporal distribution of disease symptoms and provide objective data to feed into disease forecasting models (Gao et al., 2020). The potential of these technologies lies in the increasing accuracy and reliability of sensor and other technologies, which increasingly match and surpass farmers' traditional decision-making processes (Nikolaidis, 2008). Proximal sensing systems acquire information by placing electronic instruments in contact with or in close proximity to the surface of the 'sensed object' (Zubler and Yoon, 2020). They are a subset of the so-called remote sensing technologies, which involve rapid, non-invasive analysis of plants without making physical contact with them. Proximal technologies include handheld, fixed, robotic or tractor-mounted devices. Other technologies are based on satellite, airborne or unmanned aerial vehicle (UAV) platforms (Oerke, 2020).

Global production of fresh and processed vegetables is increasing because they are such an excellent source of fibre, vitamins, antioxidants and other health-promoting compounds. They are cultivated both in the ground (open

http://dx.doi.org/10.19103/AS.2021.0095.06

fields and/or greenhouses/polytunnels) and increasingly in innovative soil-less systems, including hydroponics, aquaponics and indoor growing chambers with artificial lighting. Vegetable crops are subject to various fungal diseases exacerbated by the intensive exploitation of soils, continuous cropping and plant production in warm, humid environments that are conducive to the development of diseases. Pathogenic diseases have a serious economic impact on the sector, causing significant reductions in yield, together with high levels of food waste by reducing the aesthetic appearance, nutritional quality and shelf-life of horticultural products. Appropriate and effective phytopathological management is crucial for the competitiveness of the industry and for consumer safety.

The interest in more sustainable production systems, where plants are grown with reduced use of synthetic pesticides in favour of eco-friendly alternatives, is currently the main driver of innovation in plant protection strategies. Scientific research is continuously developing new tools for plant disease control and for the improvement of soil health, which avoid the use of synthetic chemistry. This includes new environmentally benign plant protection molecules and new pathogen-resistant and/or tolerant varieties (Juroszek et al., 2020).

Digital technologies have stimulated a growing interest in the detection of the health status of horticultural crops as part of decision support systems designed to increase the precision of disease prevention and treatment (Lázaro et al., 2020). By extending the sensing capacity of the grower both in space (i.e. monitoring the phytopathological status of field crops on a large scale) and in time (i.e. early identification of outbreaks and the capacity for short- and medium-term forecasting to drive targeted interventions), electronic devices contribute to improving the efficient use of resources and the success of disease management strategies.

Sensors act in two main ways to identify diseases:

- sensing significant changes that occur when plants are attacked by a pathogen; and
- detecting the occurrence of climatic conditions favourable to pathogenesis at the micro-environmental level.

When pathogens attack host plants, they produce a number of effects, ranging from early physiological changes, the appearance of specific and/or general symptoms of disease, loss of vigour, a reduced growth rate, and, in some cases, plant death. All of these effects can be captured by, for example, optoelectronic and olfactory devices that may allow the early and non-destructive diagnosis of plant disease. This can inform the most suitable intervention strategy (i.e. type and rate of the active ingredient to be used, and targeted application in the areas affected). In the second approach, specific micro-environmental conditions linked mainly to temperature, humidity and free water on the foliage are necessary to complete the first phases of pathogenesis, such as propagule germination, hyphal growth and host penetration. Continuous monitoring of these conditions can help to forecast future infection risks and plan targeted treatments, which can be more effective than traditional scheduled management using fungicides.

Innovative digital technologies that can be deployed include:

- sensors for the detection of micro-environmental parameters in the field;
- electronic noses; and
- imaging and optical sensors.

These sensors can then be linked among them to equip a network of sensors functional to the internet of things and artificial intelligence, to forecasting and decision-making models (Nawaz et al., 2020). These technologies are changing approaches to phytosanitary control, as well to the genetic improvement of vegetable crops for pathogen resistance characteristics, through their incorporation into high-tech phenotyping platforms (Ilakiya et al., 2020).

Key issues and challenges that are being faced in this subject, however, still remains a concern, in particular: (i) exchanging knowledge between the research world and farmers about how sensors work, what they measure and which information coming from them is expendable in plant disease management; (ii) make digital technologies and sensors easier available for growers, reducing costs, dimensions and application difficulties; (iii) gain high effective accuracy of the monitoring methods, increasing the value of the information and reduce false alerts; (iv) increase cost-efficient farm internet connections, software use, digitalization and professional background of operators.

2 Optoelectronic devices for detecting disease in vegetable plants

Optoelectronic proximal sensors work by acquiring reflected and/or emitted energy from plants, taking into account both the spatial distribution of the signals (imaging) and their site-specific higher resolution value (non-imaging). They provide a non-destructive contactless diagnosis of impaired plant status due to any adverse interaction with pathogens.

When a leaf surface is hit by an external radiation source, part of that radiation (called reflectance) is reflected back towards the external environment (Fig. 1). Reflectance is the measure of the ability of an object to reflect part of the incident radiation. It is a dimensionless quantity expressed as the ratio between the intensity of the reflected radiation and the intensity of the incident radiation. The set of electromagnetic waves of different lengths that are

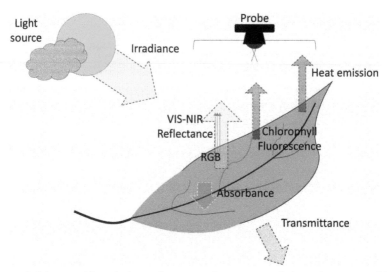

Figure 1 Scheme of how light irradiance can be partitioned through a targeted leaf among VIS-NIR spectrum reflectance, including RGB, chlorophyll fluorescence, heat emission, absorbance and transmittance and which light/energy net-losses can be recorded by the optoelectronic probes.

reflected represents the reflectance spectrum. Information from this spectrum is captured by spectral sensors. The electromagnetic spectrum can be split into different spectral regions including visible and near-mid-long infrared spectra. The detectable limits of the spectrum are defined by the type, technological setting and analytical capacity of currently available optoelectronic sensors. Imaging involves processing numerical data related to images acquired using optical sensors and the subsequent visualization of the results. These are associated with variations in physicochemical characteristics such as colour, thermographic temperature and chlorophyll fluorescence, which represent symptoms of pathogen attack.

Hyperspectral sensors are able to cover wide regions of the electromagnetic spectrum, which can include visible and near-infrared wavelengths in the range 400–3000 nm, according to the power of single or combined instruments, for example, those also using non-imaging technologies, such as spectroradiometers and cameras. Imaging hyperspectral sensors are able to capture the spectroscopic signature of a plant's status (i.e. the trace reflectance value for each measurable wavelength), including health and disease symptoms (Lowe et al., 2017). Prabhakar et al. (2013), for example, radiometry to identify leaves with different levels of yellow mosaic virus symptoms. The hyperspectral probe acquires information continuously in the designated spectrum range, with a pre-set resolution level given by the minimum step between each measured wavelength.

Hyperspectral technology involves the acquisition of a set of wavelengths *per* measuring point in non-imaging sensors. A hyperspectral image obtained by a camera with a well-defined width and height is a hypercubic dataset in which each pixel has the corresponding series of reflectance values. In order to reduce data redundancy, data processing simplifies the broad numeric array into the most significant spectral wavelengths (i.e. a few bands or narrow spectral intervals and/or their combinations). Hyperspectral analyses use vegetative indices related to a plethora of physiological (i.e. photosynthesis, water relations, plant vitality), morphological-structural (i.e. leaf area index (LAI), green biomass, growth rate, soil coverage) and biochemical traits (i.e. degree of pigmentation related to the content of chlorophyll, anthocyanins, carotenoids) (Haboudane et al., 2004; Xue and Su, 2017). Vegetative indices can be used to identify diseases and/or to highlight correlations between causal factors (pathogen) and physiological damage (symptoms) (Fig. 2). Zhao et al. (2016), for example, related the spatial distribution of angular leaf spot disease symptoms of cucumber through hyperspectral mapping of chlorophyll (*chl*) in leaves using the vegetative index *Chl-HSI*. An alternative is to use *de novo* hyperspectral algorithms extracted from reflectance measurements and machine-learning models comparing diseased and healthy samples, which can also be used for reliable detection and identification of phytopathologies in plants (Mahlein et al., 2013).

Many studies have shown the use of reflectance-derived information for the (early) detection of both direct and indirect symptoms caused by

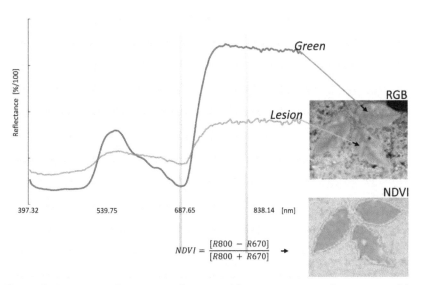

Figure 2 Hyperspectral signature of grey mold spots on tomato leaves caused by *Botrytis cinerea*, in comparison with healty leaf spectrum as, also, highlighted by RGB and NDVI images.

pathogen attacks, for example, those produced by *Corynespora cassicola* and *Xanthomonas perforans* on tomatoes (Abdulridha et al., 2020a) and *Rhizoctonia solani* on sugarbeet (Barreto et al., 2020). Bienkowski et al. (2019) used data from potato foliage reflectance spectra (Blue Wave Spectrometer 350–1150 nm), combined with supervised training of back propagation neural networks, to identify pre-symptomatic samples affected by a set of different diseases. Gold et al. (2020) detected early infections of late blight and early blight in potatoes just 1 day after inoculation by using a contact leaf spectroradiometer (350–2500 nm) and identified normalized difference spectral indices using partial least squares discriminant analysis. Hyperspectral technology was able to identify the *Arabidopsis thaliana/Sclerotinia sclerotiorum* pathosystem only 3 h after artificial inoculation of plants (Liang et al., 2019).

Hyperspectral fingerprinting of plant disease symptoms has been carried out in laboratory settings as a basis for possible in-field sensor applications. As an example, machine learning narrowed a 380–1023 nm dataset down to just five wavelengths with the potential as a non-invasive method to identify early blight and late blight diseases on tomato leaves (Xie et al., 2015). The study of reflectance data from the canopy of *Sclerotinia*-infected celery indicated the red, near-infrared, and blue wavelengths as the most significant raw spectral bands for disease prediction (Huang and Apan, 2006). Investigation of signatures for each stage of powdery mildew infection of squash, scanned by benchtop hyperspectral imaging system (range of 380–1030 nm, 281 channels, resolution of 2.1 nm) identified the most significant bands to monitor the development of the disease (Abdulridha et al., 2020b). Manganiello et al. (2021) identified high-performing hyperspectral vegetation indices in the 400–1000 nm spectral range, very useful to track the biocontrol activity of *Trichoderma* spp. against soil-borne diseases in baby-leaf vegetables (Fig. 3). A short-wavelength infrared (SWIR) spectral imaging system (950–1650 nm) has been used to extract key wavelengths identifying *Burkholderia cepacia* infection of onions (Wang et al., 2012b). Lu et al. (2018) used a line scanning spectrograph to investigate the reflectance spectra and absolute reflectance difference spectra in the wavelength range of 500–1000 nm; these were used to create multispectral images at 560 575 nm and 720 nm with the potential to detect tomato yellow leaf curl virus infection in the field. High-resolution probes used in fundamental research can generate new knowledge useful in developing cheaper, more practical spectral-based sensors, such as multispectral sensors, for automated diagnosis and/or monitoring of plant diseases in the field.

Multispectral sensors are able to work in well-defined spectral bands (a typical limit is 25) to simplify subsequent data processing. They consist of a hybrid camera system that can be directly associated via software to plant disease vegetation indices and reveal changes in plant physiology attributable to the pathogen. A five-channel camera working on the RGB and red-edge (RE) bands

Figure 3 Hyperspectral snapshot camera during an image acquiring stage on vegetable cropping system to tentatively capture disease symptoms.

was able to separate late blight infected potato plants from healthy ones using the vegetation indices SR Clgreen, RI, TCARI, TCARI/OSAVI-2, ClRed-Edge, and RE NDVI (Fernández et al., 2020). Radiometric evaluations using single-waveband reflectance values and selected vegetative indices have been used to assess the severity of *Cercospora beticola* necrosis in sugar beet (Steddom et al., 2005). A multispectral imaging system based on two synthetic narrowband indices has been developed for disease diagnosis of *Pyrenopeziza brassicae* on *Brassica napus* (Veys et al., 2019). A linear regression model showed that NIR and RE bands can be used to identify grey mold leaf infection in tomatoes caused by *Botrytis cinerea* as early as 9 h after infection (Fahrentrapp et al., 2019). With the simplification of probe numbers and complexity, multispectral sensors are often used in phenotyping platforms (Svensgaard et al., 2014), on tractor-based crop scanners for on-the-go observations (Bourgeon et al., 2016) and in the development of more accurate, low-cost and portable multispectral optical devices for disease detection (Kitić et al., 2019; Habibullah et al., 2020; Wang et al., 2020).

Lower-cost optical sensors for identifying vegetable diseases include a digital camera, which uses red-green-blue (RGB) channels analysed by software to extract useful information about changes in plant appearance. The cost-effectiveness and compactness of RGB cameras encourage their use in many applications in disease monitoring. They can be seen as simplified multispectral cameras able to detect information from three wavelengths in the spectrum visible range, red (700.47 nm), green (546.09 nm) and blue (435.79 nm) signals. This technology can identify disease symptoms based on changes in colour of affected plant organs (Padmavathi and Thangadurai, 2016). In the case of multispectral scanners, a number of RGB vegetation indices have been developed to predict plant traits to indicate possible disease (McKinnon and Hoff, 2017; Lussem et al., 2018). Artificial intelligence algorithms can be used to identify visual symptoms of the disease from the analysis of coloured images (Camargo and Smith, 2009). Image processing and pattern recognition methods are used to identify colour variations crucial for disease symptom recognition (Zhang et al., 2015). These rely on the representativeness of sample collections showing all possible shades related to possible disease symptoms. Johannes et al. (2017) have developed an image-processing algorithm based on detecting hot spots in wheat that allows recognition of diseases by different devices. Mohanty et al. (2016) have developed accurate image classifiers for smartphone-assisted disease diagnosis.

Thermal sensors are used to detect infrared energy emitted from an object in the long infrared (IR) region with wavelengths ranging from 8000 nm to 14 000 nm. Infrared energy is converted to thermographic temperature points and displayed as an image showing temperature scattering (pixel by pixel) known as a thermogram. Patterns of heat on a plant surface can be related to cooling/warming dynamics. Thermal variations in a plant are mainly driven by changes in the surrounding environment (i.e. solar radiation, air temperature, relative humidity). Thermal variations in the plant are determined by stomatal conductance regulating O_2, CO_2 and vapour exchanges (transpiration and respiration rate), as well as relative water content, physiological sap flow, cellular and tissue vitality.

Thermal temperature shifts are dependent on environmental conditions, biotic damage type and how and when the transpiration rate is affected. For example, Oerke et al., (2006) observed different thermal imaging patterns in cucumber leaves infected with downy mildew and exposed to different conditions of humidity and air temperature. In the early stages of *Pseudoperonospora cubensis* infection, areas of swelling on cucumber leaves were visible in thermograms as cooled areas; as the disease progressed, these areas increased in temperature (Lindenthal et al., 2005). In this way, thermograms can be used to identify disease symptoms. Thermal imaging is non-specific but potentially effective in early recognition of the disease, depending on the degree of resolution (the lowest differential of temperature detectable) of the camera. Wang et al. (2012a) used an IR camera, able to detect in the range 7500–13 000 nm, for non-destructive detection of pre-symptomatic leaf symptoms of Fusarium wilt in cucumber. The camera was able to detect symptoms 5 days after inoculation with a thermographic temperature difference of about 3.4°C. Tomato cucumber mosaic virus infections were observable 3 days after infection by a thermal imaging detection system equipped with a 0.06°C sensitive camera (Zhu et al., 2018). Thermal imaging can show the spatial distribution of symptoms, giving additional information on the dynamic of pathogenesis, for example, as shown in pre-symptomatic detection of potato late blight (Prashar and Jones, 2014). Thermal analysis has been used to estimate the virulence of different species causing leaf blight (belonging to the genus *Alternaria*) on oilseed rape leaves (Baranowski et al., 2015).

As in the case of hyperspectral indices, there are thermal long-wave infrared stress indices relating to air, canopy or leaf temperatures in dry or wet plants associated with general physiological parameters, which can be used to identify disease symptoms (Pineda et al., 2021). A crop water-stress index, calculated by computing temperature values using discriminant and logistic regression analysis, can identify *Rhizoctonia*-diseased lettuce plants (Sandmann et al., 2018). Rispail and Rubiales (2015) found a significant correlation between the surface leaf temperature of pea varieties and their susceptibility to *Fusarium. oxysporum f. sp. pisi*. Thermal imaging has also been proposed to assess seed viability by observing variations in heat flow and can be used to separate viable and non-viable seeds (Men et al., 2017).

Chlorophyll fluorescence imaging and non-imaging sensors capture fluorescence emissions from photochemical processes related to photosynthesis. Plants use photosynthesis to produce organic carbon biomass. The process, through photosystems I and II, entraps light energy into ATP molecules. Under light stimuli, leaves dissipate excess energy by emitting fluorescence in the red and far-red spectral range with two peaks at 685 nm and 740 nm. The majority of fluorescence is emitted by photosystem II. Fluorescence measurements carried out under different types of incident light modulation provide indicators of plant photochemical status that can be used to discriminate between healthy and infected plants (Rolfe and Scholes, 2010; Pérez-Bueno et al., 2016).

Advances in technology, such as fluorometers, have simplified the detection and visualization of chlorophyll fluorescence. Hand-held camera devices collecting fluorescence data can be used to monitor physiological changes in photosynthetic efficiency, identifying damage caused by plant disease affecting photosynthetic active leaf surfaces by limiting water and nutrient use (Gorbe and Calatayut, 2012). Disease symptoms that are produced, for example, *Bremia lactucae* in lettuce (Bauriegel et al., 2014), *Ralstonia solanacearum* in tomatoes (Kim et al., 2019) and *Uromyces appendiculatus*, *Phaeoisariopsis griseola* and *Colletotrichum lindemuthianum* in beans (Bassanezi et al., 2002), can be highlighted with pixel-based analyses of Fv/Fm images. Chlorophyll fluorescence imaging has been used for the early detection of tracheofusariosis (caused by *F. oxysporum*) in cucumber (Zhou et al., 2020). Fusarium infections in tomatoes have been identified by analysing the quantum yield of photosystem II (Wagner et al., 2006). Pineda et al. (2017) used artificial neural networks, logistic regression analyses and support vector machines trained with a set of imaging data, based on the fluorescence ratio F520/F680, to distinguish healthy fruit from those affected by soft rot and leaves affected by powdery mildew. Fluorescence quenching analysis allowed the early diagnosis of *Pseudomonas syringae* pv. *Phaseolicola* infections in bean (Rodríguez-Moreno et al., 2008).

3 Sensors for the detection of micro-environmental parameters related to disease outbreaks

This category includes all those sensors that allow the detection of microclimatic parameters in the cultivation environment. Sensors measure parameters such as temperature, humidity, wind speed, and free water which may promote pathogen spread. They can also measure parameters such as solar radiation, light intensity, water potential, wind direction and speed, and amount of rain which affect plant vigour and thus susceptibility to pathogenic infection. There is a wide range of devices, most recently using wireless technology with continuous data acquisition and remote data management (rather than a direct connection to a server). The sensors can be applied both in the environment in which the crop is developing, for monitoring soil, air and water conditions (extra-canopy), and at the plant level (intra-canopy) based on the concept of a 'plant sentinel' using single plants to estimate conditions for the surrounding crop (Fig. 4).

The information acquired by the network of sensors can be processed by algorithms based on well-calibrated forecasting models to identify threshold values related to the risk of infection. A forecast of the risk of disease, for example, enables a farmer to adjust the climate within a greenhouse, preventing thermal fluctuations and/or water levels/humidity that might promote disease. Using environmental data as inputs, these systems estimate the risk of disease occurring and alert growers when conditions might become favourable to pathogenic attacks. Models need to be calibrated and validated according to both the crop and the pathogen to ensure accurate prediction of disease risk. For example, downy mildew (*Phytophthora infestans*), a key disease in many vegetables, relies on the activity of oomycetes and are themselves highly

Figure 4 Compact climatic station for continuous monitoring of environmental parameters close to tomato plants.

dependent on environmental factors, which can be monitored and modelled. The sporulation of *Bremia lactucae* on lettuce, for example, occurs at the optimum temperature of 15°C and increases markedly with relative humidity values higher than 90%. Infection of rocket with *Hyaloperonospora parasitica* has been shown to occur at temperatures between 10°C and 26°C and relative humidity of 85%.

Electronic noses are another type of device that has been proposed for rapid and non-destructive detection of plant disease. Infected plants release a distinctive set of volatile organic compounds (VOCs) as a reaction to biotic stressors as part of plant airborne defence signalling or pathogen-associated molecular patterns (Heil and Ton, 2008). The so-called electronic noses (eNose) are devices capable of acquiring real-time information about the chemical and physical nature of gas emitted by plants including VOCs, effectively smelling the scent of plant diseases (Wilson, 2013). The eNose works by detecting variations in electrical conductivity related to the presence of gas and/or other volatile compounds in the atmosphere. This kind of device allows non-destructive, continuously acquired and early (though non-specific) diagnosis of plant diseases (Wilson, 2020; Cellini et al., 2017). eNose technologies have been proposed for the vegetable sector, for example, to identify damage by pests (Laothawornkitkul et al., 2008), bacterial infections in potatoes (Biondi et al., 2014), and grey mould and powdery mildew in tomato (Sun et al., 2016; Ghaffari et al., 2012). However, despite their potential, eNose devices need further development before they can be widely used in the horticultural sector (Sharifi et al., 2018).

4 Case study: digital and mechatronic applications on baby leaf vegetable quality chain

Baby leaf vegetables, including wild rocket, baby lettuce and baby spinach, and more other minor herbaceous species, are developed under intensive greenhouse systems in order to supply a minimally processed salad chain. The attention of consumers for ready-to-eat fresh vegetable foods leads to the preference of products from reduced or no pesticide-based cultivations. However, these conditions entail high soil-environment exploitation and contribute to making crops more susceptible to soil-borne pathogens, increasing phytopathological risks and requirements for effective disease management practices alternative to synthetic fungicides. Targeted use of resources and productive factors such as the plant protection means is focused by this case study[1] aimed at implementing advanced digital and mechatronic applications for quality horticultural supply chains, such as those of greenhouse baby leaf vegetables. In particular, research integrated digital supporting systems (i.e. hyperspectral and thermal imaging, non-imaging spectral probe and disease risk alert devices) in order to enhance performances of sustainable disease management tools (Manganiello et al., 2021) and identify digital-based variables associated to levels of disease on the specific crops (Santonicola et al., 2019) and weed infestation (Pallottino et al., 2020).

Greenhouse climatic station, equipped with temperature, humidity, foliar wetting (Fig. 5), photosynthetically active radiation, and pyranometre sensors, served to continuously monitor crop and elaborate threshold values conducive to the development of major diseases. On the other hand, spectral and thermal optoelectronic probes are applied in different growing conditions in order to find, by imaging and non-imaging elaborations, new vegetation indices and/or by artificial intelligence models able to discriminate symptomatic from healthy plants (Fig. 6).

5 Conclusion

In the next decades, horticultural production will face challenges such as climate change, loss of soil fertility and the need to feed an increasing world population. The increasing threat of disease could dramatically undermine the resilience of production systems that are currently transitioning towards more agroecological methods of cultivation driven by the growing focus on environmental protection and the production of healthy and safe food. The concept of 'producing more with less' which underpins these systems will

1 The current research is carried out in the framework of the project AgriDigit-Agrofiliere (Integrated digital technologies for the sustainable strengthening of agro-food production and transformation, financed by the Italian Ministry of Agriculture, Food and Forestry Policies (Mipaaf), grant DM 36503.7305.2018 of 20/12/2018).

Figure 5 Electronic leaf to assess foliar wetting on wild rocket cultivation.

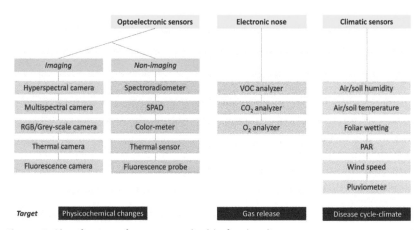

Figure 6 Classification of sensors applicable for the disease management support in vegetable cropping systems.

involve enhancing the efficacy of disease control methods by targeting their application more efficiently using digital and other technologies.

The proximal sensors discussed in this chapter can be integrated into small hand-held instruments, mounted in fixed locations, embedded in operating machines moving through the crop or carried by light drones flying just above crops (Pirna and Lache, 2010). Production systems involving proximal sensors assume the digitalization of increasing amounts of the primary production chain, giving operators the ability to continuously assess and improve methods of cultivation. A key area for development is to increase accuracy in predicting disease risk. It will be essential to develop a holistic approach. Concepts such as 'internet of things', 'artificial intelligence', 'forecasting' and 'decision-making', 'modelling', 'robotic' and 'mechatronic' will need to be integrated with concepts such as agroecology, biological control and low input practices to achieve sustainable improvements in production. Machine vision technologies based on analysis of observed parameters using vegetation indices will, for example, allow the targeted application of bio-protectants and biopesticides to replace reliance on fungicides (Ampatzidis et al., 2017).

In time, proximal sensing may facilitate the transition to a fully 'smart' horticulture. Machine learning is an important element in this transition. It can be applied to different aspects of plant disease management: identification/detection, classification, quantification and prediction through big data mining and the possibility of self-calibration with sensors as dynamic, self-regulating systems (Yang and Guo, 2017). Deep learning of large data sets using pattern recognition approaches will improve these various aspects of disease management (Ferentinos, 2018) though there are challenges linked to plant-disease management (Barbedo, 2018).

6 Where to look for further information

Specifically, research in proximal sensing applied at vegetable cropping systems could evolve by producing digital support to the crop disease management in a wider range of host/pathogen systems. The optimum would be to have substantial and significant information on how sensors work regarding the whole disease spectrum of each vegetable crop. As parallel step, the automatized acquisition of measurements and their high throughput processing giving on accurate outputs would be pursued with more perseverance. Indeed, research on this topic moves on balance between discovering new electronic technologies, such as sensors, hardware components/software, and, on the other hand, finding out field applicative solutions. Therefore, the continuous knowledge exchanging among all the involved scientists (engineers, computer scientists, physicians, agronomists, biologists, etc.) is crucial to reach successful, affordable and reliable results. However, in order to plan future research aimed at making available knowledge, technology and usefulness in horticultural crop health status detection, following a stepwise approach that take in account the system complexity is necessary. In perspective, innovation work in proximal sensing for vegetable crops will focus on the reduced size and costs of the sensors, easy to apply under the cultivation conditions (i.e. wireless, resistant to adverse working conditions, durable, energetically autonomous, etc.) and to connect with the on farm or on cloud control unit. On the other hand, high-performing software for big data management needs to be coupled with high-significant system biology models and/or algorithm to increase the accuracy of the previsions or of the interpretations. All of these technological aspects have to be defined and developed in the course of specific lab to field-experimentations paying very much attention to check the true link between the acquired data and the observed biological phenomena, which, in this specific case, are involved in disease occurrence and/or symptom expression. Because big data management is very time-consuming stage of the research, and sometime the bottleneck is difficult to overcome, the scientific solidity of the dataset has greatest importance in the flow-work. The most interesting discovering are possible when different experts from the universities, research centres, developing start-up worldwide, etc., are able to dialogue with each other constructively. This scientific field takes advantage of real multidisciplinarity.

7 References

Abdulridha, J., Ampatzidis, Y., Kakarla, S. C. and Roberts, P. (2020a). Detection of target spot and bacterial spot diseases in tomato using UAV-based and bench-top-based hyperspectral imaging techniques, *Precis. Agric.* 21, 955–978.

Abdulridha, J., Ampatzidis, Y., Roberts, P. and Kakarla, S. C. (2020b). Detecting powdery mildew disease in squash at different stages using UAV-based hyperspectral imaging and artificial intelligence, *Biosyst. Eng.* 197, 135–148.

Ampatzidis, Y., De Bellis, L. and Luvisi, A. (2017). iPathology: robotic applications and management of plants and plant diseases, *Sustainability* 9(6), 1010. https://doi.org /10.3390/su9061010.

Baranowski, P., Jedryczka, M., Mazurek, W., Babula-Skowronska, D., Siedliska, A. and Kaczmarek, J. (2015). Hyperspectral and thermal imaging of oilseed rape (*Brassica napus*) response to fungal species of the genus *Alternaria*, *PLoS ONE* 10(3), e0122913.

Barbedo, J. G. A. (2018). Factors influencing the use of deep learning for plant disease recognition, *Biosyst. Eng.* 172, 84–91.

Barreto, A., Paulus, S., Varrelmann, M. and Mahlein, A. K. (2020). Hyperspectral imaging of symptoms induced by *Rhizoctonia solani* in sugar beet: comparison of input data and different machine learning algorithms, *J. Plant Dis. Prot.* 127(4), 441–451.

Bassanezi, R. B., Amorim, L., Bergamin Filho, A. B. and Berger, R. D. (2002). Gas exchange and emission of chlorophyll fluorescence during the monocycle of rust, angular leaf spot and anthracnose on bean leaves as a function of their trophic characteristics, *J. Phytopathol.* 150(1), 37–47.

Bauriegel, E., Brabandt, H., Gärber, U. and Herppich, W. B. (2014). Chlorophyll fluorescence imaging to facilitate breeding of *Bremia lactucae*-resistant lettuce cultivars, *Comput. Electron. Agr.* 105, 74–82.

Bienkowski, D., Aitkenhead, M. J., Lees, A. K., Gallagher, C. and Neilson, R. (2019). Detection and differentiation between potato (*Solanum tuberosum*) diseases using calibration models trained with non-imaging spectrometry data, *Comput. Electron. Agric.* 167, 105056.

Biondi, E., Blasioli, S., Galeone, A., Spinelli, F., Cellini, A., Lucchese, C. and Braschi, I. (2014). Detection of potato brown rot and ring rot by electronic nose: from laboratory to real scale, *Talanta* 129, 422–430.

Bourgeon, M. A., Paoli, J. N., Jones, G., Villette, S. and Gée, C. (2016). Field radiometric calibration of a multispectral on-the-go sensor dedicated to the characterization of vineyard foliage, *Comput. Electron. Agric.* 123, 184–194.

Camargo, A. and Smith, J. S. (2009). An image-processing based algorithm to automatically identify plant disease visual symptoms, *Biosyst. Eng.* 102(1), 9–21.

Cellini, A., Blasioli, S., Biondi, E., Bertaccini, A., Braschi, I. and Spinelli, F. (2017). Potential applications and limitations of electronic nose devices for plant disease diagnosis, *Sensors (Basel)* 17(11), 2596.

Fahrentrapp, J., Ria, F., Geilhausen, M. and Panassiti, B. (2019). Detection of gray mold leaf infections prior to visual symptom appearance using a five-band multispectral sensor, *Front. Plant Sci.* 10, 628.

Ferentinos, K. P. (2018). Deep learning models for plant disease detection and diagnosis, *Comput. Electron. Agr.* 145, 311–318.

Fernández, C. I., Leblon, B., Haddadi, A., Wang, J. and Wang, K. (2020). Potato late blight detection at the leaf and canopy level using hyperspectral data, *Can. J. Remote Sens.* 46(4), 390–413.

Gao, Z., Luo, Z., Zhang, W., Lv, Z. and Xu, Y. (2020). Deep learning application in plant stress imaging: a review, *AgriEngineering* 2(3), 430–446.

Ghaffari, R., Laothawornkitkul, J., Iliescu, D., Hines, E., Leeson, M., Napier, R., More, J. P., Paul, N. D., Hewitt, C. N. and Taylor, J. E. (2012). Plant pest and disease diagnosis: electronic nose and support vector machine approach, *J. Plant Dis. Prot.* 119(5–6), 200–207.

Gold, K. M., Townsend, P. A., Chlus, A., Herrmann, I., Couture, J. J., Larson, E. R. and Gevens, A. J. (2020). Hyperspectral measurements enable pre-symptomatic detection and differentiation of contrasting physiological effects of late blight and early blight in potato, *Remote Sens.* 12(2), 286.

Gorbe, E. and Calatayut, A. (2012). Applications of chlorophyll fluorescence imaging technique in horticultural research: a review, *Sci. Hort.* 138, 24–35.

Habibullah, M., Mohebian, M. R., Soolanayakanahally, R., Bahar, A. N., Vail, S., Wahid, K. A. and Dinh, A. (2020). Low-cost multispectral sensor array for determining leaf nitrogen status, *Nitrogen* 1(1), 67–80.

Haboudane, D., Miller, J. R., Pattey, E., Zarco-Tejada, P. J. and Strachan, I. B. (2004). Hyperspectral vegetation indices and novel algorithms for predicting green LAI of crop canopies: modeling and validation in the context of precision agriculture, *Remote Sens. Environ.* 90, 337–352.

Heil, M. and Ton, J. (2008). Long-distance signalling in plant defence, *Trends Plant Sci.* 13(6), 264–272.

Huang, J. F. and Apan, A. (2006). Detection of *Sclerotinia* rot disease on celery using hyperspectral data and partial least squares regression, *J. Spat. Sci.* 51(2), 129-142.

Ilakiya, T., Parameswari, E., Davamani, V., Swetha, D. and Prakash, E. (2020). High-throughput crop phenotyping in vegetable crops, *Pharm. Innov. J.* 9(8), 184-191.

Johannes, A., Picon, A., Alvarez-Gila, A., Echazarra, J., Rodriguez-Vaamonde, S., Navajas, A. D. and Ortiz-Barredo, A. (2017). Automatic plant disease diagnosis using mobile capture devices, applied on a wheat use case, *Comput. Electron. Agr.* 138, 200-209.

Juroszek, P., Racca, P., Link, S., Farhumand, J. and Kleinhenz, B. (2020). Overview on the review articles published during the past 30 years relating to the potential climate change effects on plant pathogens and crop disease risks, *Plant Pathol.* 69(2), 179-193.

Kim, J. H., Bhandari, S. R., Chae, S. Y., Cho, M. C. and Lee, J. G. (2019). Application of maximum quantum yield, a parameter of chlorophyll fluorescence, for early determination of bacterial wilt in tomato seedlings, *Hortic. Environ. Biotechnol.* 60(6), 821-829.

Kitić, G., Tagarakis, A., Cselyuszka, N., Panić, M., Birgermajer, S., Sakulski, D. and Matović, J. (2019). A new low-cost portable multispectral optical device for precise plant status assessment, *Comput. Electron. Agric.* 162, 300-308.

Laothawornkitkul, J., Moore, J. P., Taylor, J. E., Possell, M., Gibson, T. D., Hewitt, C. N. and Paul, N. D. (2008). Discrimination of plant volatile signatures by an electronic nose: a potential technology for plant pest and disease monitoring, *Environ. Sci. Technol.* 42(22), 8433-8439.

Lázaro, E., Makowski, D., Martínez-Minaya, J. and Vicent, A. (2020). Comparison of frequentist and bayesian meta-analysis models for assessing the efficacy of decision support systems in reducing fungal disease incidence, *Agronomy* 10(4), 560.

Liang, J., Li, X., Zhu, P., Xu, N. and He, Y. (2019). Hyperspectral reflectance imaging combined with multivariate analysis for diagnosis of *Sclerotinia* stem rot on *Arabidopsis thaliana* leaves, *Appl. Sci.* 9(10), 2092.

Lindenthal, M., Steiner, U., Dehne, H. W. and Oerke, E. C. (2005). Effect of downy mildew development on transpiration of cucumber leaves visualized by digital infrared thermography, *Phytopathology* 95(3), 233-240.

Lowe, A., Harrison, N. and French, A. P. (2017). Hyperspectral image analysis techniques for the detection and classification of the early onset of plant disease and stress, *Plant Methods* 13, 80.

Lu, J., Zhou, M., Gao, Y. and Jiang, H. (2018). Using hyperspectral imaging to discriminate yellow leaf curl disease in tomato leaves, *Precis. Agric.* 19, 1-16.

Lussem, U., Bolten, A., Gnyp, M. L., Jasper, J. and Bareth, G. (2018). Evaluation of RGB-based vegetation indices from UAV imagery to estimate forage yield in grassland, *Int. Arch. Photogramm. Remote Sens. Spatial Inf. Sci.* XLII-3, 1215-1219.

Mahlein, A.-K., Rumpf, T., Welke, P., Dehne, H.-W., Plümer, L., Steiner, U. and Oerke, E.-C. (2013). Development of spectral indices for detecting and identifying plant diseases, *Remote Sens. Environ.* 128, 21-30.

Manganiello, G., Nicastro, N., Caputo, M., Zaccardelli, M., Cardi, T. and Pane, C. (2021). Functional hyperspectral imaging by high-related vegetation indices to track the wide-spectrum *Trichoderma* biocontrol activity against soil-borne diseases of baby-leaf vegetables, *Front. Plant Sci.* 12, 630059.

McKinnon, T. and Hoff, P. (2017). Comparing RGB-Based Vegetation Indices with NDVI for Drone Based Agricultural Sensing. AgribotixLLC AGBX021-17, pp. 1–8.

Men, S., Yan, L., Liu, J., Qian, H. and Luo, Q. (2017). A classification method for seed viability assessment with infrared thermography, *Sensors (Basel)* 17(4), 845.

Mohanty, S. P., Hughes, D. P. and Salathé, M. (2016). Using deep learning for image-based plant disease detection, *Front. Plant Sci.* 7, 1419.

Nawaz, M. A., Khan, T., Rasool, R. M., Kausar, M., Usman, A., Bukht, T. F. N., Ahmad, R. and Ahmad, J. (2020). Plant disease detection using internet of thing (IoT), *Int. J. Adv. Comput. Sci. Appl.* 11, 505–509.

Nikolaidis, I. (2008). Sensor networks and the law of accelerating returns, *IEEE Network* 4, 2–3.

Oerke, E. C. (2020). Remote sensing of diseases, *Annu. Rev. Phytopathol.* 58, 225–252.

Oerke, E. C., Steiner, U., Dehne, H. W. and Lindenthal, M. (2006). Thermal imaging of cucumber leaves affected by downy mildew and environmental conditions, *J. Exp. Bot.* 57(9), 2121–2132.

Padmavathi, K. and Thangadurai, K. (2016). Implementation of RGB and grayscale images in plant leaves disease detection–comparative study, *Indian J. Sci. Technol.* 9(6), 1–6.

Pallottino, F., Pane, C., Figorilli, S., Pentangelo, A., Antonucci, F. and Costa, C. (2020). Greenhouse application of light-drone imaging technology for assessing weeds severity occurring on baby-leaf red lettuce beds approaching fresh-cutting, *Span. J. Agr. Res.* 18, e0207.

Pérez-Bueno, M. L., Pineda, M., Cabeza, F. M. and Barón, M. (2016). Multicolor fluorescence imaging as a candidate for disease detection in plant phenotyping, *Front. Plant Sci.* 7, 1790.

Pineda, M., Barón, M. and Pérez-Bueno, M.-L. (2021). Thermal imaging for plant stress detection and phenotyping, *Remote Sens.* 13, 68.

Pineda, M., Pérez-Bueno, M. L., Paredes, V. and Barón, M. (2017). Use of multicolour fluorescence imaging for diagnosis of bacterial and fungal infection on zucchini by implementing machine learning, *Funct. Plant Biol.* 44(6), 563–572.

Pirna, C. and Lache, S. (2010). A new mechatronic approach for underlieing decisional processes in precision agriculture, *Bull. Transil.* 3, 27–32.

Prabhakar, M., Prasad, Y. G., Desai, S., Thirupathi, M., Gopika, K., Rao, G. R. and Venkateswarlu, B. (2013). Hyperspectral remote sensing of yellow mosaic severity and associated pigment losses in *Vigna mungo* using multinomial logistic regression models, *Crop Prot.* 45, 132–140.

Prashar, A. and Jones, H. G. (2014). Infra-red thermography as a high-throughput tool for field phenotyping, *Agronomy* 4(3), 397–417.

Rispail, N. and Rubiales, D. (2015). Rapid and efficient estimation of pea resistance to the soil-borne pathogen *Fusarium oxysporum* by infrared imaging, *Sensors (Basel)* 15(2), 3988–4000.

Rodríguez-Moreno, L., Pineda, M., Soukupová, J., Macho, A. P., Beuzón, C. R., Barón, M. and Ramos, C. (2008). Early detection of bean infection by *Pseudomonas syringae* in asymptomatic leaf areas using chlorophyll fluorescence imaging, *Photosynth. Res.* 96(1), 27–35.

Rolfe, S. A. and Scholes, J. D. (2010). Chlorophyll fluorescence imaging of plant-pathogen interactions, *Protoplasma* 247(3–4), 163–175.

Sandmann, M., Grosch, R. and Graefe, J. (2018). The use of features from fluorescence, thermography, and NDVI imaging to detect biotic stress in lettuce, *Plant Dis.* 102(6), 1101–1107.

Santonicola, L., Villecco, D., Pentangelo, A. and Pane, C. (2019). Detecting downy mildew symptoms on wild rocket leaves by hyperspectral imaging. Proceedings of the XXV National Congress of Italian Phytopathological Society, Milan, Italy, 16-18 September 2019, 131.

Sharifi, R., Lee, S. M. and Ryu, C. M. (2018). Microbe-induced plant volatiles, *New Phytol.* 220(3), 684–691.

Steddom, K., Bredehoeft, M. W., Khan, M. and Rush, C. M. (2005). Comparison of visual and multispectral radiometric disease evaluations of Cercospora leaf spot of sugar beet, *Plant Dis.* 89(2), 153–158.

Sun, Y., Wang, J. and Cheng, S. (2016). Predicting. attacked time of tomato seedling by e-nose based on kernel principal component analysis. Proceedings of the ASABE Annual International Meeting, Orlando, FL, USA, 17-20 July 2016.

Svensgaard, J., Roitsch, T. and Christensen, S. (2014). Development of a mobile multispectral imaging platform for precise field phenotyping, *Agronomy* 4(3), 322–336.

Veys, C., Chatziavgerinos, F., Al Suwaidi, A., Hibbert, J., Hansen, M., Bernotas, G., Smith, M., Yin, H., Rolfe, S. and Grieve, B. (2019). Multispectral imaging for presymptomatic analysis of light leaf spot in oilseed rape, *Plant Methods* 15, 4.

Wagner, A., Michalek, W. and Jamiolkowska, A. (2006). Chlorophyll fluorescence measurements as indicators of fusariosis severity in tomato plants, *Agron. Res.* 4, 461–464.

Wang, L., Duan, Y., Zhang, L., Wang, J., Li, Y. and Jin, J. (2020). LeafScope: a portable high-resolution multispectral imager for *in vivo* imaging soybean leaf, *Sensors (Basel)* 20(8), 2194.

Wang, M., Ling, N., Dong, X., Zhu, Y., Shen, Q. and Guo, S. (2012a). Thermographic visualization of leaf response in cucumber plants infected with the soil-borne pathogen *Fusarium oxysporum* f. sp. *cucumerinum*, *Plant Physiol. Biochem.* 61, 153–161.

Wang, W., Li, C., Tollner, E. W., Gitaitis, R. D. and Rains, G. C. (2012b). Shortwave infrared hyperspectral imaging for detecting sour skin (*Burkholderia cepacia*)-infected onions, *J. Food Eng.* 109(1), 38–48.

Wilson, A. D. (2013). Diverse applications of electronic-nose technologies in agriculture and forestry, *Sensors (Basel)* 13(2), 2295–2348.

Wilson, A. D. (2020). Non-invasive early disease diagnosis by electronic-nose and related VOC-detection devices, *Biosensors* 10(7), 73.

Xie, C., Shao, Y., Li, X. and He, Y. (2015). Detection of early blight and late blight diseases on tomato leaves using hyperspectral imaging, *Sci. Rep.* 5, 16564.

Xue, J. and Su, B. (2017). Significant remote sensing vegetation indices: a review of developments and applications, *J. Sens.* 2017, 1–17.

Yang, X. and Guo, T. (2017). Machine learning in plant disease research, *Eur. J. Bio. Med Res.* 3(1), 6–9.

Zhang, S. W., Shang, Y. J. and Wang, L. (2015). Plant disease recognition based on plant leaf image, *J. Anim. Plant Sci.* 25, 42–44.

Zhao, Y. R., Li, X., Yu, K. Q., Cheng, F. and He, Y. (2016). Hyperspectral imaging for determining pigment contents in cucumber leaves in response to angular leaf spot disease. *Sci. Rep.*, 6, 27790.

Zhou, C., Mao, J., Zhao, H., Rao, Z. and Zhang, B. (2020). Monitoring and predicting Fusarium wilt disease in cucumbers based on quantitative analysis of kinetic imaging of chlorophyll fluorescence, *Appl. Opt.* 59(29), 9118–9125.

Zhu, W., Chen, H., Ciechanowska, I. and Spaner, D. (2018). Application of infrared thermal imaging for the rapid diagnosis of crop disease, *IFAC Papers OnLine* 51(17), 424–430.

Zubler, A. V. and Yoon, J. Y. (2020). Proximal methods for plant stress detection using optical sensors and machine learning, *Biosensors* 10(12), 193.

Chapter 3

Advances in using proximal spectroscopic sensors to assess soil health

Kenneth A. Sudduth and Kristen S. Veum, USDA-ARS, USA

1 Introduction

Due to the sensitivity of the soil ecosystem and its role in global carbon cycling, multiple national and international initiatives are focused on soil carbon sequestration, soil carbon markets, and monitoring of soil health across soils and climates. Assessment of soil health involves determining how well a soil is performing its biological, chemical, and physical functions by measuring soil properties referred to as soil health indicators. Applied soil health research is focused on the needs of 'on-farm' soil health assessment and interpretation and requires tools and strategies that are affordable and accessible. The high cost and labor requirements of most currently available soil health assessments render them impractical for large-scale efforts. Ideally, an on-farm soil health assessment would be easy to measure, applicable to field conditions, affordable, and capture a wide range of soil functions that vary spatially and temporally (Doran and Parkin, 1996; Andrews et al., 2004; Karlen et al., 2017). Thus, development of rapid, low-cost methods for in-field assessment of soil health is a priority. Soil spectroscopy techniques offer high-density spatial and temporal soil

http://dx.doi.org/10.19103/AS.2022.0107.10

Published by Burleigh Dodds Science Publishing Limited, 2023.

information to help achieve agronomic and sustainability goals related to soil health. This chapter highlights the successful application of soil spectroscopy to soil health assessment, identifies novel tools and strategies that are advancing the science of soil health assessment through soil spectroscopy, and outlines areas for future research.

2 Soil spectroscopy methods

Diffuse reflectance spectroscopy (DRS) is a method that has been successfully used to estimate soil organic carbon (SOC), soil organic matter (SOM), and numerous other properties related to soil health (see reviews by Malley et al., 2004; Viscarra Rossel et al., 2006; Stenberg et al., 2010). Spectral analysis of soils is generally carried out in the visible (Vis; 400–700 nm), near-infrared [NIR; 750–2500 nm (13500–4000 cm^{-1})], or mid-infrared [MIR; 2500–23 500 nm (4000–450 cm^{-1})] range of the electromagnetic spectrum (Soriano Disla et al., 2014; Nocita et al., 2015; Johnson et al., 2019). The DRS method is based on the diffusion of incoming radiation in soil prior to being reflected and detected. This interaction of light with soil constituents produces spectral features that can subsequently be used to develop chemometric models for estimation of soil properties (Malley et al., 2004).

Spectra are generally quantified in per cent reflectance or decimal reflectance (R) as a function of wavelength. Often, reflectance will be transformed to apparent absorbance: $A = \log_{10}(1/R)$ as that quantity is more linearly related to the concentration of an absorbing compound. The fundamental vibrations of chemical bonds excited by radiation occur in the MIR region. However, instrumentation to quantify MIR DRS is more expensive, less robust to field deployment, and generally requires additional sample pretreatment beyond that required for Vis-NIR DRS. Therefore, most soil spectral analysis occurs in the NIR or Vis-NIR spectral range where overtones of the fundamental vibrations exist but become progressively weaker at shorter wavelengths. These overtones are broad and overlap with one another, such that few well-defined spectral features are found in the Vis-NIR region. This renders the spectra difficult to interpret directly because they are not amenable to assignment or quantification as individual peaks. However, Vis-NIR spectra still contain useful information on organic and inorganic soil constituents. Absorptions in the Vis region are often associated with iron-containing minerals (Sherman and Waite, 1985), and due to the inherent 'darkness' of SOM, it can also broadly affect Vis absorbance. In the NIR region, absorptions result from a variety of bonds containing carbon, nitrogen, hydrogen, and oxygen atoms, making quantification of organic compounds, clay minerals, and water possible (Fig. 1).

Published by Burleigh Dodds Science Publishing Limited, 2023.

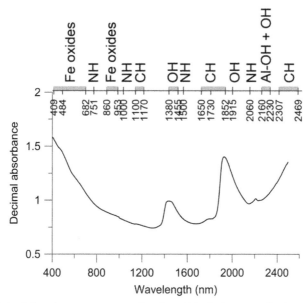

Figure 1 Typical absorbance spectrum of moist soil, including specific absorption bands for soil chemical bonds (after Katuwal et al., 2018).

2.1 Calibration approaches

Early approaches to statistical analysis of spectral datasets often involved a multiple, stepwise regression approach to select informative wavebands from the spectrum (Hruschka, 1987). This generally followed one or more spectral pretreatments to remove interference and/or linearize the relationship between spectral data and the response variable. First and second derivatives of the spectra and ratios of reflectance or absorbance at selected bands were common. Presently, the wavelength selection approach has generally been replaced by the full-spectrum approaches described later. An exception is where wavelengths are being selected with the intention of developing a discrete-waveband sensor, which could be expected to be less complex and more rugged than a full-spectrum instrument (Lee et al, 2009; Zhou et al., 2022).

The full-spectrum calibration uses multivariate techniques such as partial least squares regression (PLSR) or principal components regression (PCR; Viscarra Rossel and Behrens, 2010) that can handle highly correlated, high-dimensional data where the number of predictor variables is greater than the number of observations (Haaland and Thomas, 1988). Also, as signals in the Vis-NIR region are weaker than vibrations in the MIR region, multivariate techniques such as PLSR can overcome these challenges (Viscarra Rossel and Behrens, 2010). More recently, Veum et al. (2018) applied Bayesian covariate-assisted techniques to estimation of SOC, total nitrogen, and soil texture (i.e. clay, silt, and sand content). In addition, a range of mathematical pretreatments or spectral transformations may be required, with no consensus on best practices in the literature (Stenberg et al., 2010). Some have combined PCR or PLSR with feature selection methods to retain only the most informative spectral variables for calibration. For example, Zhou et al. (2022) used backward interval PLSR (Zou et al., 2007) to select the most informative spectral regions for estimating multiple soil properties. They reported improved accuracy compared to standard PLSR analysis for some, but not all, variables.

Various machine learning approaches have also been used to develop prediction models, allowing better representation of nonlinearities than PCR and PLSR. Some of the machine learning techniques include neural networks (Zhou et al., 2022), random forests (Viscarra Rossel and Behrens, 2010), and boosted regression trees (Brown et al., 2006). Pei et al. (2019) compared PLSR, neural networks, and random forests for estimating 11 soil physical and chemical properties. They reported that best results were obtained by PLSR in six cases, by neural networks in three cases, and by random forests in two cases. When considering multiple studies comparing calibration methods, no single algorithm, including PLSR, neural networks, or other machine learning algorithms, has consistently provided best results (Igne et al., 2010; Pei et al., 2019; Viscarra Rossel and Behrens, 2010).

Another approach to spectral estimation of soil properties relies on soil spectral libraries. These libraries have been created at a variety of scales, ranging from national (e.g. Wijewardane et al., 2016b) to global (Viscarra Rossel et al, 2016). The libraries contain soil property data, spectra, and associated metadata, which can be used with various modeling methods to estimate soil properties for spectra that are not in the library. However, these models often perform poorly at a local scale (i.e. field or farm), even when the libraries are inclusive of sites from the locality or have been augmented or 'spiked' with local samples (Wetterlind and Stenberg, 2010). Methods to improve local representations using soil spectral libraries are a current research topic. For example, Shen et al. (2022) used a process called 'deep transfer learning' to use a global soil spectral library (Viscarra Rossel et al., 2016) for estimating field-scale variations in soil carbon, finding that the approach provided improved results compared to locally calibrated models.

2.2 Lab-based soil spectroscopy

Early soil spectroscopy investigations were laboratory-based, due to the complexity and cost of spectrophotometers. Generally, soils were obtained in the field, air- or oven-dried, crushed to pass a sieve (usually 2 mm), and then scanned. A variety of complex, expensive spectrophotometers were used, but over time the portable Vis-NIR (350–2500 nm) ASD FieldSpec spectrometer (Analytical Spectral Devices, now a part of Malvern Panalytical, Malvern, UK) became almost a de-facto standard in soil research. A standard protocol for lab-based spectroscopy was proposed by Viscarra Rossel et al. (2016).

2.3 Field systems

In contrast to more traditional laboratory approaches, on-the-go sensor technology has the potential to provide high-resolution, in situ data quickly at low cost (Hummel et al., 1996). In situ sensing can also provide measurements at a high spatial or temporal density at relatively low cost (Adamchuk et al., 2004), leading to greater overall accuracy in mapping soil variability. Lab-based analyses have two sources of error: (1) analysis error due to sub-sampling and analytical determination and (2) sampling error due to point-to-point variation in soils. With traditional soil testing, analysis error is expected to be relatively low, but sampling error can be substantial due to limited sampling density across space and time. Sensors can provide a spatiotemporal sampling intensity several orders of magnitude greater than traditional methods, thus reducing sampling error and potentially reducing overall error even if analysis errors are higher.

The process of collecting soil samples in the field and sample preparation for lab-based DRS requires soil disturbance and considerable labor and processing time. These issues are especially important when dense spatial sampling is desired, for example, to estimate soil health across fields and landscapes. Thus, in situ DRS instruments have been developed to estimate soil properties in real time after the required calibrations are established. Both continuous spectrum and discrete spectral band instruments have been the subject of research and development. In the continuous spectrum approach, Sudduth and Hummel (1993) designed a mobile NIR spectrophotometer (Fig. 2) and tested it in the laboratory and field. Christy (2008) developed a mobile spectrometer-based system with a range of 920–1718 nm and used it to map spatial differences in SOM content. Mouazen et al. (2007, 2014) developed a soil sensing device with a measuring range of 306–1711 nm and reported results for estimation of total carbon, soil moisture content, pH, and phosphorus.

In-field soil sensors based on discrete spectral bands have typically used light-emitting diodes, laser diodes, or tungsten halogen lamps as light sources (An et al., 2014; Zhou et al., 2019). Compared with a spectrometer that acquires data for hundreds to thousands of spectral bands, soil sensors based on discrete spectral bands are relatively inexpensive. Shonk et al. (1991) developed a prototype real-time SOM sensor that used red light-emitting diodes as the light source. An et al. (2014) developed a total nitrogen sensor based on six discrete

Figure 2 Mobile near-infrared (1700–2420 nm) spectrophotometer developed in the late 1980s for estimating in-field variations in soil organic matter.

NIR wavebands, and Zhou et al. (2019) developed an in situ total nitrogen and soil moisture sensor with eight discrete NIR bands. Kweon and Maxton (2013) reported on a commercialized two-band (660 nm and 940 nm) sensor designed to estimate SOM, showing good agreement with laboratory-measured SOM across six test fields. Although these discrete-band sensors could be effective in estimating a single soil property, the critical sensing wavelengths for multiple soil properties are generally not coincident (Lee et al., 2009; Zhou et al., 2022). Therefore, the continuous-spectrum approach is likely more appropriate when the goal is to estimate multiple soil properties for a soil health assessment.

2.4 Moisture and other interfering factors

Prediction of multiple soil properties using DRS has been successful on air-dried surface samples (Sudduth and Hummel, 1993; Chang et al., 2001; Veum et al., 2015b), air-dried whole-profile soil samples (Lee et al., 2009), and moist samples (Sudduth and Hummel, 1993; Morgan et al., 2009; Veum et al., 2015b) scanned in the laboratory. Some studies have compared predictions obtained at different soil moisture levels, with some reporting better results with dry soil (Sudduth and Hummel, 1993; Stevens et al., 2006) and others reporting better results with moist soil (Fystro, 2002; Nocita et al., 2013). It is known that Vis-NIR spectra are particularly sensitive to soil moisture, decreasing the estimation accuracy of Vis-NIR collected in the field. In particular, hydroxyl bands from soil moisture mask some spectral features produced by SOC (Mouazen et al., 2007). It has been suggested that including a wide range of soil moisture contents in calibration models may mitigate problems associated with soil moisture variation (Sudduth and Hummel, 1993). Moreover, in a field setting, spectra are sensitive to other environmental conditions (e.g. temperature and soil structure) further decreasing prediction accuracy and the ultimate utility of spectra collected in the field (Sudduth and Hummel, 1993; Mouazen et al., 2007; Morgan et al., 2009; Reeves, 2010; Minasny et al., 2011; Ji et al., 2015).

Various techniques have been applied to account for moisture and other environmental factors to improve model performance, including external parameter orthogonalization (EPO), direct standardization, and global moisture modeling (Wijewardane et al., 2016a). The EPO algorithm removes variation due to external factors by projecting the soil spectra orthogonal to the space of unwanted variation (Roger et al., 2003). Studies have successfully applied EPO for the estimation of soil properties, including soil carbon (Minasny et al., 2011; Ge et al., 2014; Veum et al., 2018) and clay content (Ge et al., 2014; Ackerson et al., 2017; Veum et al., 2018). Alternatively, the direct standardization approach derives a transfer matrix to characterize differences between corresponding field and laboratory spectra and has successfully been used to predict SOM using a portable spectrometer (Ji et al., 2015). With the global moisture modeling technique, a secondary variable with a relationship to the primary variable is intentionally manipulated, resulting in a more robust calibration model (Kawano et al., 1995). This approach, akin to spiking, has been applied to datasets that span large geographical regions or use combined spectral libraries to estimate soil carbon and clay content (Brown, 2007; Wetterlind and Stenberg, 2010). The ultimate goal is to leverage existing spectral libraries collected on dry, laboratory-processed soils for prediction of soil properties from spectra collected in situ under variable environmental conditions.

2.5 Soil profile information

Soil health research has focused on surface soils (~0-15 cm) that are critical in plant growth and development and represent the zone of primary agricultural management. Surface soils are the interface with the atmosphere, and alteration of the upper soil horizons through management can impact many biological, physical, and chemical soil properties and processes. In addition, there are practical limitations in soil sampling equipment and budget constraints that result in an emphasis on surface soil horizons in most soil health assessments. However, vertical stratification and anisotropy in the soil profile affect infiltration, storage, and movement of water (Bouma et al., 1977), nutrient availability (Jobbágy and Jackson, 2001), carbon storage (Harrison et al., 2011), and microbial activity (Fontaine et al., 2007). For example, the subsurface argillic horizon characteristic of claypan soils impacts water partitioning, runoff, lateral movement of water through the soil, and microbial community structure and function (Hsiao et al., 2018).

Vertical stratification of soil health indicators is frequently observed under no-till systems. Adoption of no-till is one of the most important strategies for Conservation Agriculture, and it can improve soil health (Nunes et al., 2018; Veum et al., 2014). However, long-term no-till systems managed without other conservation practices (e.g. diversified cropping systems and permanent soil cover by mulching and/or cover crops) can also result in strong stratification of chemical, physical, and biological soil health indicators, with a high concentration of nutrients and SOC within the uppermost topsoil layer (~0-8 cm) and low nutrient availability and high soil compaction below that depth (Deubel et al., 2011; Houx et al., 2011; Nunes et al., 2019a; Powlson and Jenkinson, 1981). These characteristics may restrict plant root growth, decrease the absorption and translocation of water and nutrients to plants, and ultimately limit crop yield. Furthermore, soil compaction also reduces infiltration capacity, leading to increased runoff, soil erosion, and transport of pesticides and nutrients. Therefore, in-field, profile estimation of soil properties could provide vertical spatial data to inform soil health status and site-specific constraints (e.g. subsurface soil compaction) that are essential for informed management and enhanced crop production.

Previous laboratory studies have demonstrated successful estimation of profile soil properties with DRS. For example, Dalal and Henry (1986) estimated profile SOC, total nitrogen, and soil moisture with $R^2 > 0.84$. Others used NIR DRS to estimate profile soil moisture ($R^2 = 0.87$) and SOC ($R^2 > 0.77$) in the laboratory (Hummel et al., 2001), and to estimate several profile soil properties, including clay, calcium, cation exchange capacity, and SOC with $R^2 \geq 0.80$ (Lee et al., 2009). Another approach has been in-field spectrometry on intact, extracted soil cores (Kusumo, et al., 2008; Viscarra Rossel, et al., 2017). With this approach, SOC was successfully predicted at depth resolutions as small as 1 cm (Hedley et al., 2015; Roudier et al., 2015).

Acquisition of in situ profile soil information has been investigated by several groups, as it offers many potential benefits over laboratory analysis of profile samples in terms of efficiency, timeliness, and expense. A penetrometer foreoptic developed for a prototype NIR spectrophotometer estimated soil moisture with $R^2 = 0.90$ (Hummel et al., 2004). In field studies, prototype penetrometer foreoptics coupled to commercial spectrometers estimated clay content with 20-25% greater error compared to laboratory spectra on dried and sieved samples (Poggio et al., 2017; Ackerson et al., 2017). A commercial penetrometer instrument, the Veris P4000 (Veris Technologies, Salina, Kansas, USA) can be deployed in the field for profile VNIR DRS data collection (Fig. 3). Data from this instrument have been used to estimate multiple soil properties. In one study, the cross-validation R^2 of bulk density, SOC, moisture, clay, silt,

Figure 3 Veris P4000 instrument capable of sensing soil spectra (343-2202 nm), soil apparent electrical conductivity, and penetration resistance to a depth of 1 m.

and sand content were found to be 0.32, 0.67, 0.40, 0.65, 0.61, and 0.38, respectively (Cho et al., 2017b). In another study, soil texture was estimated with errors of ~6% for clay and silt, 10–11% for sand, and SOM with errors of 0.3-0.5% (Wetterlind et al., 2015). The SOC estimates from in situ Veris P4000 data were less accurate (R^2 from 0.78 to 0.90) than estimates from P4000 data obtained in the laboratory on dry soil samples (R^2 from 0.93 to 0.96; Hodge and Sudduth, 2012). Veum et al. (2018) reported similar results, with the root mean square error of prediction smaller (0.19%) for SOC estimates based on laboratory DRS than for estimates based on in situ DRS (0.26%).

2.6 Auxiliary data and sensor data fusion

Soil health reflects an integration of physical, chemical, and biological functions, and it is not currently feasible to provide a complete soil health assessment with a single soil sensing approach. In particular, models for important chemical and physical aspects of soil health have been less successful than the DRS models for biological indicators. One technique to improve estimation of inconsistent target soil health indicators is to incorporate auxiliary variables into the estimation models that may be directly or indirectly related to the target soil property or may be less expensive to measure. This can be accomplished by combining simple laboratory measurements with reflectance spectra (Brown et al., 2006; Kinoshita et al., 2012; Morgan et al., 2009; Veum et al., 2015b, 2018). Although this may overcome some of the limitations of using reflectance spectra alone, this approach still requires collection and analysis of soil in the laboratory.

Another approach combines or 'fuses' data from complementary sensors and avoids laboratory soil analysis altogether (Adamchuk et al., 2011; Cho et al., 2017b; Kusumo et al., 2008; Roudier et al., 2015; Veum et al., 2017). In one example, an in-field core-scanning system that included gamma-ray attenuation and digital imaging along with NIR spectroscopy was used to estimate multiple soil profile properties (Viscarra Rossel et al., 2017). In addition, the combination of reflectance spectra with soil apparent electrical conductivity (EC_a) and penetration resistance (or soil strength) sensor data, as implemented in the Veris P4000 instrument, has demonstrated improved estimates of multiple soil properties (Pei et al., 2019; Veum et al., 2017). These studies illustrate some potential for rapid quantification of soil health by fusing auxiliary laboratory data or data obtained from complementary sensors. However, critical gaps remain in the development and application of sensors for nutrient availability, soil structure, and advanced soil biological indicators.

3   Estimation of soil health indicators and indices

3.1 Soil biological properties

Within the developing science of soil health, soil biology is playing a leading role in advancing our understanding of ecosystem function as well as agricultural productivity and sustainability (Lehman et al., 2015). Therefore, it is critical to gain a better understanding of soil biological function across space and time. A wide range of soil biological properties usually quantified by traditional laboratory methods are potential candidates for spectroscopic estimation. These range from standard SOM and SOC measurements to carbon and nitrogen mineralization assays (e.g. Doran, 1987; Franzluebbers et al., 1996), soil enzyme assays (e.g. Dick, 1994; Eivazi and Tabatabai, 1988), a rapid carbon oxidation test (Culman et al., 2012; Weil et al., 2003), a rapid protein extraction (Hurisso et al., 2018), and analyses of the microbial community via genomics (Manter et al., 2017) or high-throughput lipid analysis (Buyer and Sasser, 2012).

DRS has been successfully used to estimate SOC or SOM for some time in the laboratory as an alternative to wet chemistry methods (see reviews by Stenberg et al., 2010; Viscarra Rossel et al., 2006). In addition to SOC and SOM, spectra in this range have been used to successfully estimate several other biological soil health indicators across a range of soils and ecosystems, including total nitrogen, β-glucosidase activity, oxidizable carbon, microbial biomass-carbon, microbial lipids, and soil respiration (Cécillon et al., 2009; Chang et al., 2001; Chaudhary et al., 2012; Kinoshita et al., 2012; Palmborg and Nordgren, 1993, 1996; Pietikäinen and Fritze, 1995; Sudduth and Hummel, 1993; Veum et al., 2014, 2015b; Zornoza et al., 2008).

Published by Burleigh Dodds Science Publishing Limited, 2023.

Despite the relative success of these studies, estimation of these soil properties may be only achieved by proxy through correlation with the spectral signature of SOC or SOM. For example, good estimates may be the result of changes in surrogate (i.e. proxy or highly correlated) soil properties instead of directly detecting changes in the property of interest when using NIR spectroscopy methods (Reeves, 2010; Vågen et al., 2006). Organic carbon bonds and inorganic mineral bonds are the primary sources of spectral features (see Cécillon et al., 2009; Viscarra Rossel et al., 2006), and when a model relies on an indirect, proxy measurement, performance is dependent upon the relationship between the target soil property and the proxy measurement.

3.2 Soil physical properties

Physical soil properties, such as bulk density, aggregate stability, and water-filled pore space, reflect multiple soil functions such as infiltration capacity, resistance to erosion, and microbial habitat (Angers, 1992; Franzluebbers, 1999; Logsdon and Karlen, 2004). In-field estimation of these important physical soil health indicators using reflectance spectra has been less consistent than for soil biological properties, although some studies have demonstrated success with soil texture and aggregate stability (Bogrekci and Lee, 2005; Chang et al., 2001; Vågen et al., 2006) and bulk density (Cho et al., 2017a).

In some cases, spectral data have been supplemented with data from other sensor variables, including soil strength (Hemmat and Adamchuk, 2008) and EC_a (Corwin and Lesch, 2005). Soil strength is highly correlated with other dynamic soil physical properties, such as bulk density, macroporosity, aggregate stability, moisture, and degree of compaction, as well as inherent soil properties such as texture (Chung et al., 2006; Nunes et al., 2019a), and root growth (Nunes et al., 2019b). While vertically operating cone penetrometers are the standard device for measuring soil strength, horizontal, on-the-go sensors have also been developed to improve data collection efficiency (Chung et al., 2006; Hemmat and Adamchuk, 2008). Soil apparent electrical conductivity is affected by numerous attributes, including soil texture, mineralogy, cation exchange capacity, and moisture (Doolittle and Brevik, 2014; McNeill, 1992; Sudduth et al., 2013). In non-saline soils, texture variations generally have the largest effect, so EC_a has been used to map soil texture differences (Anderson-Cook et al., 2002; Kitchen et al., 1996). However, the lack of a unique relationship between EC_a values and specific soil properties, mainly due to the multiple properties that can affect EC_a in any given situation, has hindered its use for quantitative assessment of soil properties in a practical setting. While good results are often obtained in research, it is difficult to elucidate protocols that will transfer those results to practice.

Published by Burleigh Dodds Science Publishing Limited, 2023.

3.3 Soil chemical properties

Chemical soil health indicators primarily reflect nutrient availability for crop growth and are typically represented by laboratory measurements of soil pH, electrical conductivity (paste), extractable phosphorus, and extractable potassium (Smith and Doran, 1996; Staggenborg et al., 2007; Wienhold et al., 2009). Precision, in-field soil health assessments could provide high-resolution information for improved fertility management decisions (e.g. right time, right place, right rate) and would enhance the sustainability of production systems by reducing environmental impacts and improving profitability. However, sensing of available phosphorus and potassium using reflectance spectra has been met with limited success. A few researchers (e.g. Bogrekci and Lee, 2005; Daniel et al., 2003) reported consistently good ($R^2 > 0.8$) results for both phosphorus and potassium, but other studies (e.g. Ge et al., 2007; Lee et al., 2009; Viscarra Rossel et al., 2006) reported consistently poor ($R^2 < 0.5$) results. This lack of consistency has been attributed to the fact that spectral estimation of phosphorus and potassium relies on the covariation of nutrient concentrations with other, optically active soil constituents (Stenberg et al., 2010). For a more accurate estimation of soil nutrients, La et al. (2016) supplemented spectral data with laboratory data from electrochemical sensors, improving phosphorus and potassium estimates to $R^2 \geq 0.95$. Additional information on the electrochemical sensing of nutrients can be found in reviews by Kim et al. (2009) and Sinfield et al. (2010).

3.4 Soil health indices

The ability to estimate multiple soil health indicators and provide a soil health score to land managers without expensive soil sampling and analysis would support multiple national and international sustainability initiatives related to carbon sequestration, carbon markets, and soil health. To this end, many studies have related Vis-NIR reflectance to individual soil health indicators as described earlier or suites of soil health measurements. In addition, Vis-NIR has been applied to the direct estimation of soil health indices. The general aim of a soil health index is to translate measured values of soil health indicators into a more user-friendly, interpretable classification or score, typically related to an outcome such as erosion, crop productivity, or water quality. A wide range of soil health scoring frameworks are available depending on the application and the outcome of interest, and Vis-NIR has been used to estimate soil health indices in ecosystems across the globe. Vis-NIR spectra have been applied by Vågen et al. (2006) to estimate a 3-category soil fertility index based on 10 soil health indicators in Madagascar, by Cohen et al. (2006) to classify 3 tiers of soil health based on 17 indicators in the USA, and by Kinoshita et al. (2012) to estimate a 3-category soil quality index for soils in Western Kenya.

Published by Burleigh Dodds Science Publishing Limited, 2023.

Few studies, however, have evaluated the simultaneous estimation of biological, physical, and chemical soil health indicators with the goal of estimating a comprehensive soil health index (as opposed to a categorical soil health index). Veum et al. (2015a) estimated soil health scores for soils from the Midwestern USA based on the Soil Management Assessment Framework (SMAF). The SMAF index scores measured values using a set of non-linear curves with thresholds for 'more is better', 'less is better', or mid-point optima (Andrews et al., 2004). Veum et al. (2015b) found that their Vis-NIR models successfully estimated the biological components of the SMAF score but performed poorly for the chemical and physical components of the SMAF scores. Due to the non-linear nature of the SMAF scoring curves, the distribution of the scores likely impacted model performance, particularly for the chemical and nutrient scores. In particular, the mid-point optima scoring curves for soil pH and phosphorus performed poorly in the models. Overall, the nonlinearity of the SMAF scoring algorithms presented a challenge for the PLSR approach. Thus, non-linear techniques may provide more robust estimation models. However, a single soil health dataset is unlikely to exhibit ideal distributional characteristics across all soil health indicators simultaneously, which may reduce the ability to estimate a comprehensive soil health index using Vis-NIR spectra alone.

Adding auxiliary variables to Vis-NIR spectra may improve model performance, particularly for estimation of a soil health index where multiple soil health indicators are represented. Veum et al. (2015b) added selected auxiliary variables, alone and in combination, to improve estimation of the SMAF soil health score. Bulk density, water-filled pore space, pH, aggregate stability, and/or phosphorus measurements all resulted in improved estimation of the overall SMAF score, suggesting that Vis-NIR sensors in conjunction with other sensors or simple field and/or laboratory analyses may improve index estimates. Selecting auxiliary variables that reflect aspects of soil health that are not represented well by Vis-NIR, such as physical, chemical, or nutrient indicators, may enhance estimation. As noted earlier, no single sensor has demonstrated the ability to estimate a comprehensive soil health index. This challenge represents an ideal opportunity for the application of sensor fusion technology. For example, Veum et al. (2017) noted that the fusion of EC_a and cone index (penetrometer) data with Vis-NIR improved model estimates of the physical SMAF category and subsequently the overall SMAF soil health score, while chemical and fertility-related soil health indicators remained poorly estimated.

Overall, current research supports using sensor fusion to improve estimation of soil health indicators and SMAF scores. Modern agriculture is challenged in attaining a balance of agronomic productivity with long-term sustainability, and sensor fusion technology along with novel data analysis techniques has the potential to provide high-resolution soil health assessment results for more informed management decisions. Current research points to the need for improved sensors for chemical and physical soil health indicators and better techniques to handle the challenges of varying environmental conditions for an effective in-field approach to soil health assessment.

4 Case study: combining spectra and auxiliary sensor data for improved soil health estimation

A series of related studies illustrate the potential and limitations of soil spectroscopy for estimating variables important in quantifying soil health. These studies focused on the Central Claypan Region in northeastern Missouri, USA. The soils found at these sites, characterized as claypan soils, are primarily of the Mexico-Putnam association (fine, montmorillonitic, mesic Udollic Ochraqualfs). These soils were formed in moderately fine textured loess over a fine textured pedisediment. Surface textures range from a silt loam to a silty clay loam. The subsoil claypan horizon(s) are silty clay loam, silty clay, or clay and may commonly contain as much as 50-60% montmorillonitic clay. This abrupt increase in clay content over a few centimeters is the distinguishing factor of claypan soils, and the strong stratification between topsoil and high-clay subsoil contributes to poor soil water holding capacity, excessive surface runoff, and in many years, insufficient soil water for optimum crop growth.

4.1 Laboratory Vis-NIR estimation of soil organic carbon

The first in this sequence of studies (Chaudhary et al., 2012) had the goal of evaluating laboratory-based Vis-NIR DRS as a tool for discriminating differences in SOC among systems in a long-term cropping systems experiment. Surface soil samples were collected from three different grain cropping systems and three perennial grass systems in the experiment. To provide an independent calibration dataset, other samples were collected from fields under various management systems located within a few kilometers of the experimental site. Spectral data were collected in the laboratory with an ASD FieldSpec Pro FR spectrometer between 400 nm and 2500 nm on both field-moist and oven-dry soil samples. Laboratory SOC data were provided by dry combustion analysis. Models estimating SOC were established with PLSR on the calibration dataset, and those models were then applied to the plot data with good results ($R^2 \geq$ 0.81; Fig. 4). Spectral analysis of oven-dry samples provided better results than with field-moist samples, with both able to detect SOC differences among cropping systems. However, neither was as successful in detecting differences as dry combustion lab data.

Published by Burleigh Dodds Science Publishing Limited, 2023.

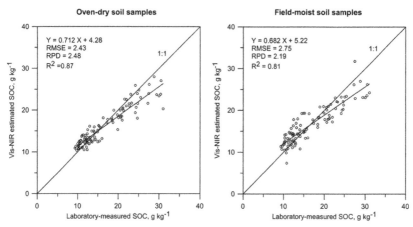

Figure 4 Visible and near-infrared (Vis-NIR) spectroscopy-estimated vs. laboratory-measured values of soil organic carbon (SOC) for oven-dry and field-moist plot soil samples (after Chaudhary et al., 2012).

4.2 Laboratory Vis-NIR estimation of soil health indicators and scoring functions

This study (Veum et al., 2015b) employed the same experimental sites and soil samples as Chaudhary et al. (2012). The goal was to evaluate laboratory Vis-NIR spectroscopy as a tool to estimate soil health indicator variables and soil health scores. The indicator variables and scoring categories were taken from those used in the SMAF (Andrews et al., 2004). This soil health index combines values of multiple indicator variables into calculated scores for biological, physical, and chemical categories, as well as an overall SMAF score. Biological variables were SOC, β-glucosidase, microbial biomass-carbon, and mineralizable nitrogen. Physical variables were bulk density, water-filled pore space, and water-stable aggregates. The chemical category included extractable phosphorus and potassium, pH, and electrical conductivity (paste).

Laboratory Vis-NIR spectroscopy was effective in estimating biological indicators but not those in the physical or chemical categories (Fig. 5). When estimating scoring functions, similar results were found, with the biological SMAF score well estimated (R^2 = 0.76) but not the scores for the other categories ($R^2 \leq 0.27$). Although the overall SMAF score was reasonably well-estimated (R^2 = 0.69), we expected this could be improved with a better estimation of physical and chemical indicators through the incorporation of data from auxiliary sensors.

4.3 Improving Vis-NIR soil health estimates with auxiliary data

To improve DRS performance for SMAF soil health scores, Veum et al. (2017) combined laboratory Vis-NIR spectral data with auxiliary data collected in the

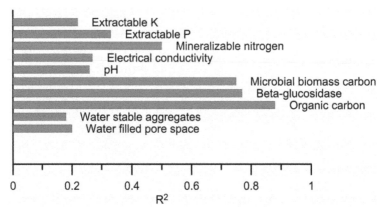

Figure 5 Accuracy of laboratory visible and near-infrared spectroscopy applied to oven-dry soil samples for estimating selected soil health indicators.

field. Soil samples were collected at 2 depths (0–5 cm and 5–15 cm) at 108 locations across a 10-ha research site including different cropping systems and landscape positions. Samples were oven-dried, and spectra (400–2500 nm) were obtained in the laboratory with an ASD FieldSpec Pro FR spectrometer. Near each sample collection point, a Veris Profiler 3000 instrument (Veris Technologies, Salina, KS, USA) was used to obtain in situ EC_a and penetration resistance data, which were then averaged over the soil sample depth increments. Models were created by PLSR with spectra alone and with spectra, EC_a, and penetration resistance data.

Results showed that adding auxiliary variables improved model estimates for all SMAF category scores and the overall SMAF score (Fig. 6). Substantial improvement was seen in the physical score, which then translated to improvement in the overall score. Improvements to biological and chemical scores were minor. While the biological SMAF score was well estimated with spectra alone (Fig. 6), the chemical score was not. Including other auxiliary data, such as that from electrochemical sensors (La et al., 2016) could be one approach to improve models for chemical scores.

4.4 Profile in situ soil sensing using Vis-NIR spectra and auxiliary data

Studies described by Cho et al. (2017b) and Pei et al. (2019) compared soil property estimates by Vis-NIR spectra alone to those from datasets that also included EC_a and penetration resistance (Fig. 7). Data were collected with a Veris P4000 instrument over multiple field sites to a depth of approximately 1 m. Spectra and auxiliary variables were averaged over depth increments corresponding with soil horizon delineations and calibration soil samples were

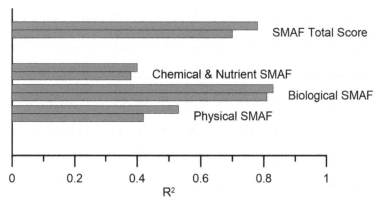

Figure 6 Improvement in Soil Management Assessment Framework (SMAF) total and categorical chemical, biological, and physical scores by including auxiliary variables of soil apparent electrical conductivity (EC_a) and penetration resistance along with Vis-NIR spectra (green bars), as compared to spectra alone (brown bars).

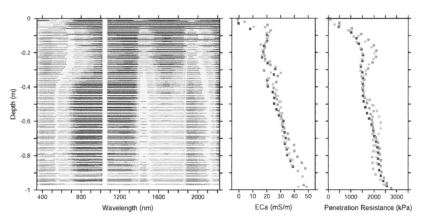

Figure 7 Soil spectra, apparent electrical conductivity (EC_a), and penetration resistance data obtained by four vertical probings to 1 m at a claypan field sample site. Spectral data are shown in absorbance units, with highest absorbance in red and lowest in blue.

obtained over those same horizons and analyzed in the laboratory. In both studies, only slight improvement was seen when adding the auxiliary variables. Pei et al. (2019) reported that root mean square error decreased by more than 5% for only 1 of the 11 soil properties evaluated. It should be noted that only three soil properties – SOC, pH, and extractable potassium – were common between these studies and the SMAF-oriented evaluations described earlier. This is due in part to the concentration of SOM and associated soil health indicators in the near-surface horizons of most soils. However, profile knowledge of keystone soil health indicators, such as SOC, and inherent soil properties, such as soil texture, can contribute to better soil management recommendations and soil health assessment interpretation.

5 Conclusion

Soil spectroscopy has the potential to provide rapid, cost-effective, high-resolution data to support critical initiatives related to soil health. Current research results support the use of Vis-NIR sensors for in-field assessment and quantification of biological soil health indicators, even under conditions of variable soil moisture content. Important soil physical or chemical soil health indicators may need to be supplied by auxiliary data collected using complementary sensors, field test kits, or simple laboratory measurements. Overall, soil spectroscopy using sensor-based technology in the field has the potential to reliably assess soil health for improved sustainability, profitability, and environmental protection.

6 Future trends in research

Achieving the goal of rapid, in-field assessment of soil health will allow for the creation of comprehensive maps with soil function layers gleaned from in-field, sensor estimates of soil health indicators. Ideally, the soil health indicator data will be transformed via on-the-go software into interpretive maps for producers along with recommendations for management. Despite the successes in this area to date, many important research and development needs remain, including new sensor technology to measure soil strength, soil nutrients, and biological characteristics directly and accurately. For example, the development of novel in-field sensors that collect real-time soil process data (e.g. soil enzyme activity) will expand our knowledge of soil variability to improve crop production and better protect soil and water resources. Integration of advanced sensor platforms and data analysis techniques will almost certainly be required. Sensor optimization to improve data quality and reliability is needed, and methods to address a range of environmental conditions will be necessary to move additional sensor techniques from the laboratory to the field.

Following data collection, improved soil health interpretation frameworks are also necessary to provide scientifically sound guidance to landowners and producers. Increased accessibility to soil health information, interpretation, and management recommendations via user-friendly software applications will allow landowners to make science-based decisions. As interpretation tools are expanded at the national and global scale, data-driven soil health management information will help optimize management decisions to provide economic and environmental benefits.

Published by Burleigh Dodds Science Publishing Limited, 2023.

7   Where to look for further information

Two books that provide thorough coverage of spectroscopy uses in agriculture are Williams and Norris (1987) and Roberts et al. (2004). Both cover spectroscopy basics, analysis methods, and applications. A chapter in the second book (Malley et al., 2004) provides coverage of soil applications. Good reviews specific to soil applications include the articles by Viscarra Rossel et al. (2006), Stenberg et al. (2010), and Nocita et al. (2015).

Current advances in soil spectroscopy are often presented at the International, European, and Asian Conferences on Precision Agriculture, Information about and links to proceedings papers from these three biennial conferences can be found at the website of the International Society of Precision Agriculture (www.ispag.org). International journals that publish soil spectroscopy research include *Biosystems Engineering, Computers and Electronics in Agriculture, European Journal of Soil Science, Geoderma, Journal of the ASABE* (previously *Transactions of the ASABE*), *Precision Agriculture,* and *Soil Science Society of America Journal.*

For details on laboratory methods for soil health analysis, a recent two-volume set published by the Soil Science Society of America serves as excellent references: Approaches to Soil Health Analysis, Volume 1 (Karlen et al., 2021a) and Laboratory Methods for Soil Health Analysis, Volume 2 (Karlen et al., 2021b).

8   References

Ackerson, J. P., Morgan, C. L. S. and Ge, Y. (2017). Penetrometer-mounted VisNIR spectroscopy: Application of EPO-PLS to in situ VisNIR spectra, *Geoderma* 286, 131-138.

Adamchuk, V. I., Hummel, J. W., Morgan, M. T. and Upadhyaya, S. K. (2004). On-the-go soil sensors for precision agriculture, *Comput. Electron. Agric.* 44(1), 71-91.

Adamchuk, V. I., Viscarra Rossel, R. A., Sudduth, K. A. and Schulze Lammers, P. (2011). Sensor fusion for precision agriculture. In: Thomas, C. (Ed.), *Sensor Fusion - Foundation and Applications.* Tech, Rijeka, Croatia, pp. 27-40.

An, X., Li, M., Zheng, L., Liu, Y. and Sun, H. (2014). A portable soil nitrogen detector based on NIRS, *Precis. Agric.* 15(1), 3-16.

Anderson-Cook, C. M., Alley, M. M., Roygard, J. K. F., Khosla, R., Noble, R. B. and Doolittle, J. A. (2002). Differentiating soil types using electromagnetic conductivity and crop yield maps, *Soil Sci. Soc. Am. J.* 66(5), 1562-1570.

Andrews, S. S., Karlen, D. L. and Cambardella, C. A. (2004). The soil management assessment framework: A quantitative soil quality evaluation method, *Soil Sci. Soc. Am. J.* 68(6), 1945-1962.

Angers, D. A. (1992). Changes in soil aggregation and organic carbon under corn and alfalfa, *Soil Sci. Soc. Am. J.* 56(4), 1244-1249.

Bogrekci, I. and Lee, W. S. (2005). Improving phosphorus sensing by eliminating soil particle size effect in spectral measurement, *Trans. ASAE* 48, 1971-1978.

Bouma, J., Jongerius, A., Boersma, O., Jager, A. and Schoonderbeek, D. (1977). The function of different types of macropores during saturated flow through four swelling soil horizons, *Soil Sci. Soc. Am. J.* 41(5), 945-950.

Brown, D. J. (2007). Using a global VNIR soil-spectral library for local soil characterization and landscape modeling in a 2nd-order Uganda watershed, *Geoderma* 140(4), 444-453.

Brown, D. J., Shepherd, K. D., Walsh, M. G., Mays, M. D. and Reinsch, T. G. (2006). Global soil characterization with VNIR diffuse reflectance spectroscopy, *Geoderma* 132(3-4), 273-290.

Buyer, J. S. and Sasser, M. (2012). High throughput phospholipid fatty acid analysis of soils, *Appl. Soil Ecol.* 61, 127-130.

Cécillon, L., Barthès, B. G., Gomez, C., Ertlen, D., Genot, V., Hedde, M., Stevens, A. and Brun, J. J. (2009). Assessment and monitoring of soil quality using near-infrared reflectance spectroscopy (NIRS), *Eur. J. Soil Sci.* 60(5), 770-784.

Chang, C. W., Laird, D. A., Mausbach, M. J. and Hurburgh, C. R. (2001). Near-infrared reflectance spectroscopy-principal components regression analysis of soil properties, *Soil Sci. Soc. Am. J.* 65(2), 480-490.

Chaudhary, V. P., Sudduth, K. A., Kitchen, N. R. and Kremer, R. J. (2012). Reflectance spectroscopy detects management and landscape differences in soil carbon and nitrogen, *Soil Sci. Soc. Am. J.* 76(2), 597-606.

Cho, Y., Sheridan, A. H., Sudduth, K. A. and Veum, K. S. (2017a). Comparison of field and laboratory VNIR spectroscopy for profile soil property estimation, *Trans. ASABE* 60(5), 1503-1510.

Cho, Y., Sudduth, K. A. and Drummond, S. T. (2017b). Profile soil property estimation using a VIS-NIR-EC-force probe, *Trans. ASABE* 60(3), 683-692.

Christy, C. D. (2008). Real-time measurement of soil attributes using on-the-go near infrared reflectance spectroscopy, *Comput. Electron. Agric.* 61(1), 10-19.

Chung, S. O., Sudduth, K. A. and Hummel, J. W. (2006). Design and validation of an on-the-go soil strength profile sensor, *Trans. ASABE* 49(1), 5-14.

Cohen, M. J., Dabral, S., Graham, W. D., Prenger, J. P. and Debusk, W. F. (2006). Evaluating ecological condition using soil biogeochemical parameters and near infrared reflectance spectra, *Environ. Monit. Assess.* 116(1-3), 427-457.

Corwin, D. L. and Lesch, S. M. (2005). Apparent soil electrical conductivity measurements in agriculture, *Comput. Electron. Agric.* 46(1-3), 11-43.

Culman, S. W., Snapp, S. S., Freeman, M. A., Schipanski, M. E., Beniston, J., Lal, R., Drinkwater, L. E., Franzluebbers, A. J., Glover, J. D., Grandy, A. S., Lee, J., Six, J., Maul, J. E., Mirksy, S. B., Spargo, J. T. and Wander, M. M. (2012). Permanganate oxidizable carbon reflects a processed soil fraction that is sensitive to management, *Soil Sci. Soc. Am. J.* 76(2), 494-504.

Dalal, R. C. and Henry, R. J. (1986). Simultaneous determination of moisture, organic carbon, and total nitrogen by near-infrared reflectance spectrophotometry, *Soil Sci. Soc. Am. J.* 50(1), 120-123.

Daniel, K. W., Tripathi, N. K. and Honda, K. (2003). Artificial neural network analysis of laboratory and in situ spectra for the estimation of macronutrients in soils of Lop Buri (Thailand), *Soil Res.* 41(1), 47-59.

Deubel, A., Hofmann, B. and Orzessek, D. (2011). Long-term effects of tillage on stratification and plant availability of phosphate and potassium in a loess chernozem, *Soil Till. Res.* 117, 85-92.

Dick, R. P. (1994). Soil enzyme activity as an indicator of soil quality. In: Doran, J. W., Coleman, D. C., Bezdicek, D. F. and Stewart, B. A. (Eds). *Defining Soil Quality for a Sustainable Environment*, Soil Science Society of America and American Society of Agronomy, Madison, WI, pp. 107–124.

Doolittle, J. A. and Brevik, E. C. (2014). The use of electromagnetic induction techniques in soils studies, *Geoderma* 223-225, 33–45.

Doran, J. W. (1987). Microbial biomass and mineralizable nitrogen distributions in no-tillage and plowed soils, *Biol. Fertil. Soils* 5(1), 68–75.

Doran, J. W. and Parkin, T. B. (1996). Quantitative indicators of soil quality: A minimum data set. In: Doran, J. W. and Jones, A. J. (Eds.), *Methods for Assessing Soil Quality*. Soil Science Society of America, Madison, WI, pp. 25–37.

Eivazi, F. and Tabatabai, M. A. (1988). Glucosidases and galactosidases in soils, *Soil Biol. Biochem.* 20(5), 601–606.

Fontaine, S., Barot, S., Barré, P., Bdioui, N., Mary, B. and Rumpel, C. (2007). Stability of organic carbon in deep soil layers controlled by fresh carbon supply, *Nature* 450(7167), 277–280.

Franzluebbers, A. J. (1999). Microbial activity in response to water-filled pore space of variably eroded southern Piedmont soils, *Appl. Soil Ecol.* 11(1), 91–101.

Franzluebbers, A. J., Haney, R. L., Hons, F. M. and Zuberer, D. A. (1996). Determination of microbial biomass and nitrogen mineralization following rewetting of dried soil, *Soil Sci. Soc. Am. J.* 60(4), 1133–1139.

Fystro, G. (2002). The prediction of C and N content and their potential mineralization in heterogeneous soil samples using VIS-NIR spectroscopy and comparative methods, *Plant Soil* 246(2), 139–149.

Ge, Y., Morgan, C. L. S. and Ackerson, J. P. (2014). VisNIR spectra of dried ground soils predict properties of soils scanned moist and intact, *Geoderma* 221-222, 61–69.

Ge, Y., Thomasson, J. A., Morgan, C. L. and Searcy, S. W. (2007). VNIR diffuse reflectance spectroscopy for agricultural soil property determination based on regression-kriging, *Trans. ASABE* 50(3), 1081–1092.

Haaland, D. M. and Thomas, E. V. (1988). Partial least-squares methods for spectral analyses. 1. Relation of other quantitative calibration methods and the extraction of qualitative information, *Anal. Chem.* 60(11), 1193–1202.

Harrison, R., Footen, P. and Strahm, B. (2011). Deep soil horizons: Contribution and importance to soil carbon pools and in assessing whole-ecosystem response to management and global change, *For. Sci.* 57, 67–76.

Hedley, C., Roudier, P. and Maddi, L. (2015). VNIR soil spectroscopy for field analysis, *Commun. Soil Sci. Plant Anal.* 46(Suppl. 1), 104–121.

Hemmat, A. and Adamchuk, V. I. (2008). Sensor systems for measuring soil compaction: Review and analysis, *Comput. Electron. Agric.* 63(2), 89–103.

Hodge, A. M. and Sudduth, K. A. (2012). Comparison of two spectrometers for profile soil carbon sensing, *American Society of Agricultural and Biological Engineers Annual International Meeting*, ASABE, St. Joseph, MI, Paper No. 121338240.

Houx, J. H., Wiebold, W. J. and Fritschi, F. B. (2011). Long-term tillage and crop rotation determines the mineral nutrient distributions of some elements in a Vertic Epiaqualf, *Soil Till. Res.* 112(1), 27–35.

Hruschka, W. R. (1987). Data analysis: Wavelength selection methods. In: Williams, P. and Norris, K. (Eds.), *Near-Infrared Technology in the Agricultural and Food Industries*, American Association of Cereal Chemists, St. Paul, MN, pp. 35–56.

Hsiao, C.-J., Sassenrath, G. F., Zeglin, L. H., Hettiarachchi, G. M. and Rice, C. W. (2018). Vertical changes of soil microbial properties in claypan soils, *Soil Biol. Biochem.* 121, 154–164.

Hummel, J. W., Ahmad, I. S., Newman, S. C., Sudduth, K. A. and Drummond, S. T. (2004). Simultaneous soil moisture and cone index measurement, *Trans. ASAE* 47, 607–618.

Hummel, J. W., Gaultney, L. D. and Sudduth, K. A. (1996). Soil property sensing for site-specific crop management, *Comput. Electron. Agric.* 14(2–3), 121–136.

Hummel, J. W., Sudduth, K. A. and Hollinger, S. E. (2001). Soil moisture and organic matter prediction of surface and subsurface soils using an NIR soil sensor, *Comput. Electron. Agric.* 32(2), 149–165.

Hurisso, T. T., Moebius-Clune, D. J., Culman, S. W., Moebius-Clune, B. N., Thies, J. E. and van Es., H. M. (2018). Soil protein as a rapid soil health indicator of potentially available organic nitrogen, *Agric. Environ. Lett.* 3, 180006.

Igne, B., Reeves, J. B., III, McCarty, G., Hively, W. D., Lund, E. and Hurburgh, C. R., Jr. (2010). Evaluation of spectral pretreatments, partial least squares, least squares support vector machines and locally weighted regression for quantitative spectroscopic analysis of soils, *J. Near Infrared Spectrosc.* 18(3), 167–176.

Ji, W., Viscarra Rossel, R. A. and Shi, Z. (2015). Accounting for the effects of water and the environment on proximally sensed vis–NIR soil spectra and their calibrations, *Eur. J. Soil Sci.* 66(3), 555–565.

Jobbágy, E. G. and Jackson, R. B. (2001). The distribution of soil nutrients with depth: Global patterns and the imprint of plants, *Biogeochemistry* 53(1), 51–77.

Johnson, J. M., Vandamme, E., Senthilkumar, K., Sila, A., Shepherd, K. D. and Saito, K. (2019). Near-infrared, mid-infrared or combined diffuse reflectance spectroscopy for assessing soil fertility in rice fields in sub-Saharan Africa, *Geoderma* 354, 113840.

Karlen, D. L., Goeser, N. J., Veum, K. S. and Yost, M. A. (2017). On-farm soil health evaluations: Challenges and opportunities, *J. Soil Water Conserv.* 72(2), 26A–31A.

Karlen, D. L., Stott, D. E. and Mikha, M. M. (Eds.). (2021a). *Approaches to Soil Health Analysis* (Vol. 1), Wiley, Hoboken, NJ.

Karlen, D. L., Stott, D. E. and Mikha, M. M. (Eds.). (2021b). *Laboratory Methods for Soil Health Analysis* (Vol. 2), Wiley, Hoboken, NJ.

Katuwal, S., Knadel, M., Moldrup, P., Norgaard, T., Greve, M. H. and deJonge, L. W. (2018). Visible-near-infrared spectroscopy can predict mass transport of dissolved chemicals through intact soil, *Sci. Rep.* 8(1), 11188.

Kawano, S., Abe, H. and Iwamoto, M. (1995). Development of a calibration equation with temperature compensation for determining the Brix value in intact peaches, *J. Near Infrared Spectrosc.* 3(4), 211–218.

Kim, H. J., Sudduth, K. A. and Hummel, J. W. (2009). Soil macronutrient sensing for precision agriculture, *J. Environ. Monit.* 11(10), 1810–1824.

Kinoshita, R., Moebius-Clune, B. N., van Es, H. M., Hively, W. D. and Bilgilis, A. V. (2012). Strategies for soil quality assessment using visible and near-infrared reflectance spectroscopy in a Western Kenya chronosequence, *Soil Sci. Soc. Am. J.* 76(5), 1776–1788.

Kitchen, N. R., Sudduth, K. A. and Drummond, S. T. (1996). Mapping of sand deposition from 1993 midwest floods with electromagnetic induction measurements, *J. Soil Water Conserv.* 51, 336–340.

Kusumo, B. H., Hedley, C. B., Hedley, M. J., Hueni, A., Tuohy, M. P. and Arnold, G. C. (2008). The use of diffuse reflectance spectroscopy for in situ carbon and nitrogen analysis of pastoral soils, *Soil Res.* 46(7), 623-635.

Kweon, G. and Maxton, C. (2013). Soil organic matter sensing with an on-the-go optical sensor, *Biosys. Eng.* 115, 66-81.

La, W. J., Sudduth, K. A., Kim, H. J. and Chung, S. O. (2016). Fusion of spectral and electrochemical sensor data for estimating soil macronutrients, *Trans. ASABE* 59(4), 787-794.

Lee, K. S., Lee, D. H., Sudduth, K. A., Chung, S. O., Kitchen, N. R. and Drummond, S. T. (2009). Wavelength identification and diffuse reflectance estimation for surface and profile soil properties, *Trans. ASAE* 52, 683-695.

Lehman, R., Cambardella, C., Stott, D., Acosta-Martinez, V., Manter, D., Buyer, J., Maul, J., Smith, J., Collins, H., Halvorson, J., Kremer, R., Lundgren, J., Ducey, T., Jin, V. and Karlen, D. (2015). Understanding and enhancing soil biological health: The solution for reversing soil degradation, *Sustainability* 7(1), 988-1027.

Logsdon, S. D. and Karlen, D. L. (2004). Bulk density as a soil quality indicator during conversion to no-tillage, *Soil Till. Res.* 78(2), 143-149.

Malley, D. F., Martin, P. D. and Ben-Dor, E. (2004). Application in analysis of soils. In: Roberts, C. A., Workman, J., Jr. and Reeves, J. B., III (Eds). Near-Infrared Spectroscopy in Agriculture, *American Society of Agronomy*. Crop Science Society of America and Soil Science Society of America, Madison, WI, pp. 729-784.

Manter, D. K., Delgado, J. A., Blackburn, H. D., Harmel, D., Pérez de León, A. A. and Honeycutt, C. W. (2017). Opinion: Why we need a national living soil repository, *Proc. Natl. Acad. Sci. U. S. A.* 114(52), 13587-13590.

McNeill, J. D. (1992). Rapid, accurate mapping of soil salinity by electromagnetic ground conductivity meters. In: Topp, G. C., Reynolds, W. D. and Green, R. E. (Eds.), *Advances in Measurement of Soil Physical Properties: Bringing Theory into Practice*, Soil Science Society of America, Madison, WI, pp. 209-229.

Minasny, B., McBratney, A. B., Bellon-Maurel, V., Roger, J. M., Gobrecht, A., Ferrand, L. and Joalland, S. (2011). Removing the effect of soil moisture from NIR diffuse reflectance spectra for the prediction of soil organic carbon, *Geoderma* 167-168, 118-124.

Morgan, C. L. S., Waiser, T. H., Brown, D. J. and Hallmark, C. T. (2009). Simulated in situ characterization of soil organic and inorganic carbon with visible near-infrared diffuse reflectance spectroscopy, *Geoderma* 151(3-4), 249-256.

Mouazen, A. M., Alhwaimel, S. A., Kuang, B. and Waine, T. (2014). Multiple on-line soil sensors and data fusion approach for delineation of water holding capacity zones for site specific irrigation, *Soil Till. Res.* 143, 95-105.

Mouazen, A. M., Maleki, M. R., De Baerdemaeker, J. and Ramon, H. (2007). On-line measurement of some selected soil properties using a VIS-NIR sensor, *Soil Till. Res.* 93(1), 13-27.

Nocita, M., Stevens, A., Noon, C. and van Wesemael, B. (2013). Prediction of soil organic carbon for different levels of soil moisture using Vis-NIR spectroscopy, *Geoderma* 199, 37-42.

Nocita, M., Stevens, A., van Wesemael, B., Aitkenhead, M., Bachmann, M., Barthès, B., Ben Dor, E., Brown, D. J., Clairotte, M., Csorba, A., Dardenne, P., Demattê, J. A. M., Genot, V., Guerrero, C., Knadel, M., Montanarella, L., Noon, C., Ramirez-Lopez, L., Robertson, J., Sakai, H., Soriano-Disla, J. M., Shepherd, K. D., Stenberg, B., Towett, E. K., Vargas, R. and Wetterlind, J. (2015). Soil spectroscopy: An alternative to wet chemistry for soil monitoring. In: Sparks, D. L. (Ed.), *Advances in Agronomy* (Vol. 132), Academic Press, Burlington, MA, pp. 139-159.

Nunes, M. R., Karlen, D. L., Denardin, J. E. and Cambardella, C. A. (2019a). Corn root and soil health indicator response to no-till production practices, *Agric. Ecosyst. Environ.* 285.

Nunes, M. R., Pauletto, E. A., Denardin, J. E., Suzuki, L. E. A. S. and van Es, H. M. (2019b). Dynamic changes in compressive properties and crop response after chisel tillage in a highly weathered soil, *Soil Till. Res.* 186, 183-190.

Nunes, M. R., van Es, H. M., Schindelbeck, R., James Ristow, A. J. and Ryan, M. (2018). No-till and cropping system diversification improve soil health and crop yield, *Geoderma* 328, 30-43.

Palmborg, C. and Nordgren, A. (1993). Modelling microbial activity and biomass in forest soil with substrate quality measured using near infrared reflectance spectroscopy, *Soil Biol. Biochem.* 25(12), 1713-1718.

Palmborg, C. and Nordgren, A. (1996). Partitioning the variation of microbial measurements in forest soils into heavy metal and substrate quality dependent parts by use of near infrared spectroscopy and multivariate statistics, *Soil Biol. Biochem.* 28(6), 711-720.

Pei, X., Sudduth, K. A., Veum, K. S. and Li, M. (2019). Improving in-situ estimation of soil profile properties using a multi-sensor probe, *Sensors (Basel)* 19(5), 1011.

Pietikäinen, J. and Fritze, H. (1995). Clear-cutting and prescribed burning in coniferous forest: Comparison of effects on soil fungal and total microbial biomass, respiration activity and nitrification, *Soil Biol. Biochem.* 27(1), 101-109.

Poggio, M., Brown, D. J. and Bricklemyer, R. S. (2017). Comparison of Vis-NIR on in situ, intact core, and dried, sieved soil to estimate clay content at field to regional scales, *Eur. J. Soil Sci.* 68(4), 434-448.

Powlson, D. S. and Jenkinson, D. S. (1981). A comparison of the organic matter, biomass, adenosine triphosphate and mineralizable nitrogen contents of ploughed and direct-drilled soils, *J. Agric. Sci.* 97(3), 713-721.

Reeves, J. B. (2010). Near- versus mid-infrared diffuse reflectance spectroscopy for soil analysis emphasizing carbon and laboratory versus on-site analysis: Where are we and what needs to be done?, *Geoderma* 158(1-2), 3-14.

Roberts, C. A., Workman, J., Jr. and Reeves, J. B., III. (2004). *Near-Infrared Spectroscopy In Agriculture*, American Society of Agronomy, Crop Science Society of America, and Soil Science Society of America, Madison, WI.

Roger, J. M., Chauchard, F. and Bellon Maurel, V. (2003). EPO-PLS external parameter orthogonalisation of PLS application to temperature-independent measurement of sugar content of intact fruits, *Chemom. Intellig Lab. Syst.* 66(2), 191-204.

Roudier, P., Hedley, C. B. and Ross, C. W. (2015). Prediction of volumetric soil organic carbon from field-moist intact soil cores, *Eur. J. Soil Sci.* 66(4), 651-660.

Shen, Z., Ramirez-Lopez, L., Behrens, T., Cui, L., Zhang, M., Walden, L., Wetterlind, J., Shi, Z., Sudduth, K. A., Baumann, P., Song, Y., Catambay, K. and Viscarra Rossel, R. A. (2022). Deep transfer learning of global spectra for local soil carbon monitoring, *ISPRS J. Photogramm.* 188, 190-200.

Sherman, D. M. and Waite, T. D. (1985). Electronic spectra of Fe^{3+} oxides and oxide hydroxides in the near IR to near UV, *Am. Mineralogist* 70, 1262-1269.

Shonk, J. L., Gaultney, L. D., Schulze, D. G. and Van, G. E. (1991). Spectroscopic sensing of soil organic matter content, *Trans. ASAE* 34, 1978-1984.

Sinfield, J. V., Fagerman, D. and Colic, O. (2010). Evaluation of sensing technologies for on-the-go detection of macronutrients in cultivated soils, *Comput. Electron. Agric.* 70(1), 1–18.

Smith, J. L. and Doran, J. W. (1996). Measurement and use of pH and electrical conductivity for soil quality analysis. In: Doran, J. W. and Jones, A. J. (Eds.), *Methods for Assessing Soil Quality*, Soil Science Society of America, Madison, WI, pp. 169–185.

Soriano Disla, J. M., Janik, L. J., Viscarra Rossel, R. A., Macdonald, L. M. and McLaughlin, M. J. (2014). The performance of visible, near-, and mid-infrared reflectance spectroscopy for prediction of soil physical, chemical, and biological properties, *Appl. Spectrosc. Rev.* 49(2), 139–186.

Staggenborg, S. A., Carignano, M. and Haag, L. (2007). Predicting soil pH and buffer pH in situ with a real-time sensor, *Agron. J.* 99(3), 854–861.

Stenberg, B., Viscarra Rossel, R. A., Mouazen, A. M. and Wetterlind, J. (2010). Visible and near infrared spectroscopy in soil science. In: Sparks, D. L. (Ed.), *Advances in Agronomy* (Vol. 107), Academic Press, Burlington, MA, pp. 163–215.

Stevens, A., van Wesemael, B., Vandenschrick, G., Touré, S. and Tychon, B. (2006). Detection of carbon stock change in agricultural soils using spectroscopic techniques, *Soil Sci. Soc. Am. J.* 70(3), 844–850.

Sudduth, K. A. and Hummel, J. W. (1993). Soil organic matter, CEC, and moisture sensing with a prototype NIR spectrometer, *Trans. ASAE* 36, 1571–1582.

Sudduth, K. A., Myers, D. B., Kitchen, N. R. and Drummond, S. T. (2013). Modeling soil electrical conductivity-depth relationships with data from proximal and penetrating ECa sensors, *Geoderma* 199, 12–21.

Vågen, T.-G., Shepherd, K. D. and Walsh, M. G. (2006). Sensing landscape level change in soil fertility following deforestation and conversion in the highlands of Madagascar using Vis-NIR spectroscopy, *Geoderma* 133(3–4), 281–294.

Veum, K. S., Goyne, K. W., Kremer, R. J., Miles, R. J. and Sudduth, K. A. (2014). Biological indicators of soil quality and soil organic matter characteristics in an agricultural management continuum, *Biogeochemistry* 117(1), 81–99.

Veum, K. S., Kremer, R. J., Sudduth, K. A., Kitchen, N. R., Lerch, R. N., Baffaut, C., Stott, D. E., Karlen, D. L. and Sadler, E. J. (2015a). Conservation effects on soil quality indicators in the Missouri Salt River Basin, *J. Soil Water Conserv.* 70(4), 232–246.

Veum, K. S., Parker, P. A., Sudduth, K. A. and H Holan, S. (2018). Predicting profile soil properties with reflectance spectra via Bayesian covariate assisted external parameter orthogonalization, *Sensors (Basel)* 18(11), 3869.

Veum, K. S., Sudduth, K. A., Kremer, R. J. and Kitchen, N. R. (2015b). Estimating a soil quality index with VNIR reflectance spectroscopy, *Soil Sci. Soc. Am. J.* 79(2), 637–649.

Veum, K. S., Sudduth, K. A., Kremer, R. J. and Kitchen, N. R. (2017). Sensor data fusion for soil health assessment, *Geoderma* 305, 53–61.

Viscarra Rossel, R. A., Behrens, T., Ben-Dor, E., Brown, D. J., Demattê, J. A. M., Shepherd, K. D., Shi, Z., Stenberg, B., Stevens, A., Adamchuk, V., Aïchi, H., Barthès, B. G., Bartholomeus, H. M., Bayer, A. D., Bernoux, M., Böttcher, K., Brodský, L., Du, C. W., Chappell, A., Fouad, Y., Genot, V., Gomez, C., Grunwald, S., Gubler, A., Guerrero, C., Hedley, C. B., Knadel, M., Morrás, H. J. M., Nocita, M., Ramirez-Lopez, L., Roudier, P., Campos, E. M. R., Sanborn, P., Sellitto, V. M., Sudduth, K. A., Rawlins, B. G., Walter, C., Winowiecki, L. A., Hong, S. Y. and Ji, W. (2016). A global spectral library to characterize the world's soil, *Earth Sci. Rev.* 155, 198–230.

Viscarra Rossel, R. A., Lobsey, C. R., Sharman, C., Flick, P. and McLachlan, G. (2017). Novel proximal sensing for monitoring soil organic C stocks and condition, *Environ. Sci. Technol.* 51(10), 5630–5641.

Viscarra Rossel, R. A., Walvoort, D. J. J., McBratney, A. B., Janik, L. J. and Skjemstad, J. O. (2006). Visible, near infrared, mid infrared or combined diffuse reflectance spectroscopy for simultaneous assessment of various soil properties, *Geoderma* 131(1–2), 59–75.

Viscarra Rossel, R. A. and Behrens, T. (2010). Using data mining to model and interpret soil diffuse reflectance spectra, *Geoderma* 158(1–2), 46–54.

Weil, R. R., Islam, K. R., Stine, M. A., Gruver, J. B. and Samson-Liebig, S. E. (2003). Estimating active carbon for soil quality assessment: A simplified method for laboratory and field use, *Am. J. Alt. Agric.* 18, 3–17.

Wetterlind, J., Piikki, K., Stenberg, B. and Söderström, M. (2015). Exploring the predictability of soil texture and organic matter content with a commercial integrated soil profiling tool, *Eur. J. Soil Sci.* 66(4), 631–638.

Wetterlind, J. and Stenberg, B. (2010). Near-infrared spectroscopy for within-field soil characterization: Small local calibrations compared with national libraries spiked with local samples, *Eur. J. Soil Sci.* 61(6), 823–843.

Wienhold, B. J., Karlen, D. L., Andrews, S. S. and Stott, D. E. (2009). Protocol for indicator scoring in the soil management assessment framework (SMAF), *Renew. Agric. Food Syst.* 24(4), 260–266.

Wijewardane, N. K., Ge, Y. and Morgan, C. L. S. (2016a). Prediction of soil organic and inorganic carbon at different moisture contents with dry ground VNIR: A comparative study of different approaches, *Eur. J. Soil Sci.* 67(5), 605–615.

Wijewardane, N. K., Ge, Y., Wills, S. and Loecke, T. (2016b). Prediction of soil carbon in the conterminous United States: Visible and near infrared reflectance spectroscopy analysis of the rapid carbon assessment project, *Soil Sci. Soc. Am. J.* 80(4), 973–982.

Williams, P. and Norris, K. (1987). *Near-Infrared Technology in the Agricultural and Food Industries*, American Association of Cereal Chemists, St. Paul, MN.

Zhou, P., Sudduth, K. A., Veum, K. S. and Li, M. (2022). Extraction of reflectance spectra features for estimation of surface, subsurface, and profile soil properties, *Comput. Electron. Agric.* 196, 106845.

Zhou, P., Zhang, Y., Yang, W., Li, M., Liu, Z. and Liu, X. (2019). Development and performance test of an in-situ soil total nitrogen-soil moisture detector based on near-infrared spectroscopy, *Comput. Electron. Agric.* 160, 51–58.

Zornoza, R., Guerrero, C., Mataix-Solera, J., Scow, K. M., Arcenegui, V. and Mataix-Beneyto, J. (2008). Near infrared spectroscopy for determination of various physical, chemical and biochemical properties in Mediterranean soils, *Soil Biol. Biochem.* 40(7), 1923–1930.

Zou, X., Zhao, J. and Li, Y. (2007). Selection of the efficient wavelength regions in FT-NIR spectroscopy for determination of SSC of "Fuji" apple based on BiPLS and FiPLS models, *Vib. Spectrosc.* 44(2), 220–227.

Chapter 4

Using remote and proximal sensor data in precision agriculture applications

Luciano S. Shiratsuchi and Franciele M. Carneiro, Louisiana State University, USA; Francielle M. Ferreira, São Paulo State University (UNESP), Brazil; Phillip Lanza and Fagner A. Rontani, Louisiana State University, USA; Armando L. Brito Filho, São Paulo State University (UNESP), Brazil; Getúlio F. Seben Junior, State University of Mato Grosso (UNEMAT), Brazil; ny N. Brandao, Brazilian Agricultural Research Corporation (EMBRAPA), Brazil; Carlos A. Silva Junior, State University of Mato Grosso (UNEMAT), Brazil; Paulo E. Teodoro, Federal University of Mato Grosso do Sul (UFMS), Brazil; and Syam Dodla, Louisiana State University, USA

1 Introduction

One of the biggest challenges in agriculture is to increase crop yields without expanding production into new areas. It has been suggested that agricultural production needs to double by 2050 to meet increasing demand (Foley et al., 2011; The Royal Society, 2016; Narvaez et al., 2017). Precision agriculture (PA) provides one potential way to improve crop yields on existing agricultural land and help ensure existing land is used more sustainably. The International Society of Precision Agriculture (ISPA) has defined PA as 'a management strategy that takes account of temporal and spatial variability to improve sustainability of agricultural production' (ISPA, 2022). By identifying varying soil and crop characteristics in a field, PA enables more precise (and potentially more sustainable) variable rate application of inputs such as fertilizers and pesticides. This is achieved by applying the most appropriate inputs at the

http://dx.doi.org/10.19103/AS.2022.0107.19

right location, rate, and time and in the right manner, often referred to as site-specific management (SSM). Due to factors such as the heterogeneity of soil and variability in crop response to changing environmental conditions, in-field variation can be significant both spatially and temporally.

Sensors are therefore a key component of PA. They allow patterns of in-field temporal and spatial variability to be identified cost effectively, which then allow more precise and targeted application of inputs (Queiroz et al., 2020). A good example is nitrogen (N) fertilization. Crops vary in their N requirements at differing stages of growth while individual plants and parts of fields may have differing N needs at differing times of the year. N fertilization management is always a challenge since the N cycle in the soil is highly dynamic and complex and varies significantly over short ranges. Management must also take account of multiple loss pathways (e.g. denitrification, volatilization, runoff, and leaching) which result in negative environmental impacts (Marchi, 2021). Variable rate N fertilization, based on measuring differences in plant nutrient status, and then establishing distinct crop management zones requiring different treatments, provides the opportunity to optimize N use efficiency and reduces both economically and environmentally costly losses (Elbl et al., 2021).

This chapter reviews key issues in using sensor data in PA and, in particular, their mode of deployment (proximal or remote). It assesses the relative strengths and weaknesses of proximal sensing techniques, compared with imaging data typically acquired from remote sensing (RS) platforms, before assessing trade-offs in sensor data resolution, as well as sources of error in the way data are processed. The chapter concludes by looking at ways of integrating remote and proximal sensor data, to utilize the beneficial characteristics of each type of data to improve the impact PA in improving efficiency and sustainability.

2 Remote and proximal sensing in agriculture

There are many different sensor principles and technologies that can be employed. However, the method of deployment has a significant impact on what sensors can measure and the value of the data obtained. Sensor deployment can be broadly classified as proximal (or terrestrial) and remote.

2.1 Proximal and terrestrial sensors

Proximal soil sensing (PSS) is a well-developed area of research and is defined as soil sensor measurements that are obtained either in contact with or within a short range (<2 m) of the soil. A key advantage of proximal sensors is their mode of operation, which can be either passive or active, non-contact or invasive. An example of the latter is soil water probes that can provide high temporal resolution and information through depth (i.e. the soil profile) but typically are limited in spatial coverage, i.e. the data obtained are typically representative of only a small volume of soil and are not broadly representative of soil across a field. As well as being deployed in situ, proximal soil sensors can also be deployed on the go, e.g. by mounting sensors on agricultural vehicles to provide better spatial coverage. However, spatial coverage may still be limited by how much of the field a vehicle covers while temporal coverage will be limited by how often a vehicle is deployed.

Crop or plant sensors can also be considered 'proximal' although terms such as 'ground-based' or 'terrestrial' are also used to distinguish these types of sensor from remote sensor systems (aerial or satellite-based – discussed later). Similarly to soil sensors, proximal plant sensor measurements can be active or passive, invasive (where plant samples are collected and processed for sensor measurement), in contact with the plant (e.g. the Soil Plant Analysis Development (SPAD) meter to measure leaf chlorophyll concentrations), or can operate within a short range of a crop (e.g. typically within a range of 2-3 m). Definitions of 'proximal' have been complicated by the development of sensors mounted on unmanned aerial vehicles (UAVs) capable of hovering a few feet above a crop, pointing to an increasing degree of overlap between proximal and remote sensor technologies in areas such as degree of spatial resolution (Ferguson, 2019).

As for PSS, plant or crop sensors can be deployed on the go by mounting sensors on agricultural vehicles such as tractors, with data collected during farming operations such as fertilizer/pesticide application or harvesting, as has shown in monitoring crops such as cotton (Schielack and Thomassen, 2016; Sui et al., 2012). Active canopy sensors (ACS), based on portable or vehicle-mounted active proximal sensors, provide real-time monitoring of crop characteristics such as nutrient (e.g. N) status. Types of active proximal sensors include:

- handheld chlorophyll meters, such as (SPAD) 502 plus (Konika Minolta® Inc., Tokyo, Japan) and LEAF (FT Green LLC®, Wilmington, DE, USA)
- handheld canopy reflectance sensors such as GreenSeeker (Trimble Inc., Sunnyvale, CA, USA), RapidSCAN CS-45 sensor (Holland Scientific® Inc., Lincoln, NE, USA) and Crop Circle Series (Holland Scientific®, Inc.)

ACS-based yield predictions conducted at the stem elongation stage of crop growth have been used for topdressing fertilizer recommendations. Active proximal sensors have now been combined with N fertilization algorithms to enable on-the-go variable N application (Colaço and Bramley, 2018). Active sensors such as GreenSeeker, OptRX, CropSpec, and N sensor can be installed on agriculture machines such as sprayers to monitor and map N status of plants and then target N application to match plant requirements. Combine harvesters equipped with yield monitors and global navigation satellite system sensors (to map position and allow navigation of vehicles) can be used to produce maps of spatial yield variability in a field while harvesting.

For both soil and plant sensing, deploying sensors proximally can provide a wide range of detailed information. It provides an opportunity to deploy a wide potential range of sensor technologies. Proximal crop sensor technologies include contact and in situ sensors (such as chlorophyll meters), ranging sensors (e.g. using acoustic or laser techniques), and electromagnetic (EM) sensors, as well as passive, active, and invasive systems (Ferguson, 2019). There is also a wide range of proximal soil monitoring techniques such as electrical, optical, temperature, electrochemical, radioactive radiation, and mechanical sensors (Gebbers, 2019). Proximal sensor technologies allow the rapid and inexpensive collection of precise, fine-resolution data on soil or crop characteristics (Viscarra Rossel et al., 2011). However, a key limitation is limited spatial information and the ability to obtain both high spatial and temporal coverage simultaneously.

2.2 Remote sensors

RS involves measuring objects from a distance without physical contact with the object being measured (e.g. aerial or orbital sensors on a drone, aircraft, or satellite). Sensors operating remotely can be deployed on aerial platforms (e.g. aircraft, unmanned aircraft systems (UASs) or UAVs (commonly known as drones) or orbital platforms (artificial satellites orbiting the Earth)) (Crepani, 1993; Jensen, 2009; Florenzano, 2011; Molin et al., 2015) (Fig. 1). As noted earlier, the development of UAVs capable of operating a few feet above a crop has blurred the distinction between remote and proximal sensors. To make it possible to map spatial variability on the scale needed for PA (e.g. mapping a

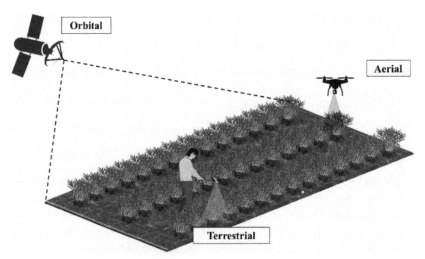

Figure 1 Collection of data using terrestrial, aerial, and orbital platforms. Source: Brito Filho (2022).

whole field), UAVs can operate at heights ranging from 30 m to 100 m or more (Zhang and Kovacs, 2019). In identifying in-field variability, PA makes particular use of RS technologies.

The two main types of aerial sensors are aircraft designed to carry sensor systems and UAS or UAVs. They can carry high-definition sensors with high spatial resolutions (potentially in centimeters or even millimeters) (Zheng et al., 2020). They can also carry cameras able to capture different wavelengths (Fig. 2) (Guo et al., 2016; Adao et al., 2017; Yuan et al., 2021). Manned aircraft have advantages such as the ability to carry a large array of heavier sensors with high spatial resolution, to provide data in real-time/near-real-time, to cover large areas (which might otherwise be difficult to access), and to map over larger scales (Yang, 2019).

With their low cost, ease of operation, and ability to fly at low altitudes, UAS have become a popular RS platform, complementing ground-based and more traditional RS platforms such as aircraft and satellites (Hunt and Daughtry, 2018; Yang, 2019). UAS-mounted sensors are capable of ultra-high spatial imagery resolution (in cm or less), making them well-suited to field scouting (Zhang and Kovacs, 2019). Speed and quality in the processing of drone imagery are developing rapidly.

A key limit to commercial UAS usage is battery life, but other constraints include the weight of instrumentation drones can carry, the distance drones can fly away from the operator (e.g. in terms of maintaining contact for control), altitude limits (typically 120 m/400 ft), and regulations restricting use (e.g. proximity to potential hazards).

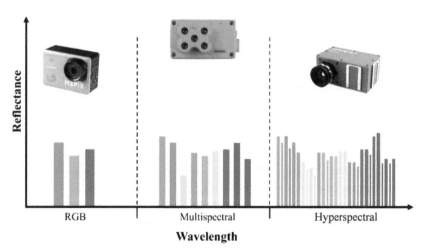

Figure 2 Different spectral resolutions as RGB (visible rays), multispectral, and hyperspectral. Source: Brito Filho (2022).

The use of orbiting satellites to gather data about the Earth itself had military origins but has now been applied in many other areas such as weather and environmental monitoring (Ge et al., 2020; Singh et al., 2020), monitoring patterns of urban development (Huang et al., 2020), and agricultural management (Tedesco et al., 2021). A key advantage is the ability to image very large areas. Agricultural applications of satellite imaging have been limited by problems such as relatively coarse spatial resolution, interruptions in coverage (depending on the speed of orbit), cloud cover, and delays in communicating data back to Earth (Mulla, 2013), though recent developments in high-resolution satellite sensors have significantly narrowed the gap in spatial resolution between traditional satellite and airborne imagery (Yang, 2019). Satellites can now carry a range of high-quality sensors able to capture images in hundreds of spectral bands and spatial resolutions that vary from kilometers to centimeters (Zanotta et al., 2019). The development of sensor technology has been accompanied by advances in mathematical modeling to analyze this data rapidly in identifying crop characteristics and the in-field variability required for PA applications (Dash et al., 2010; Santos et al., 2021).

When comparing the unique characteristics of proximal and RS, key differences are the mode of operation (e.g. active or passive), the spatial and temporal resolution of information obtained, and also the specificity of this information, i.e. does it capture variation in crop response due to some factor or does it specifically measure and quantify this factor (e.g. N status). Generally, RS can be considered most suited to addressing variability within a field, to apply treatments more precisely across the field, while proximal sensing can better attribute the cause of this variability, allowing the selection of correct treatments and rates.

3   Active and passive sensors

Most sensors used in agricultural operations are imaging sensors which analyze objects by measuring EM radiation (EMR) generated from the sun and reflected or emitted from targets on the Earth's surface (Mobley, 1994; Moreira, 2011). EMR is composed of an EM spectrum split into bands called wavelengths, each of which has its own frequency (Lorenzzetti, 2015). While some waves (gamma rays to microwave rays) are blocked by the atmosphere, others (visible light to microwave rays) reach the Earth's surface (Novo, 2008). The EMR bands most used in agriculture research are visible waves, near-infrared (NIR) waves, mid-infrared waves, and microwaves (Moreira, 2011; Molin et al., 2015). When EMR is in contact with a target (e.g. a plant), it is either absorbed, transmitted, or reflected. RS sensors typically collect reflectance data.

Plant leaf reflectance over the visible and NIR range contains useful information about crop growth and health (Fig. 3) (Jensen, 1996; Thenkabail et al., 2000; Hansen and Schjoerring, 2003; Brandão, 2009; Shiratsuchi, 2014).

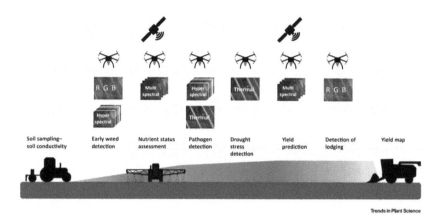

Figure 3 Application of remote sensing in agriculture. Source: Maes and Steppe (2019).

Within the visible range, leaf reflectance is lower due to light absorption by leaf pigments involved in photosynthesis, in particular chlorophylls which absorb light in the blue and red region of the spectrum (Lichtenthaler and Buschmann, 2001; Gitelson et al., 2006). In the NIR region, however, these leaf pigments do not absorb light and leaf reflectance is many times higher in magnitude (Kumar and Silva, 1973). Given this pattern of light scattering, a sharp increase between red and NIR regions can be observed in the reflectance of green parts of plants, named the red edge. Red edge reflectance has been correlated with many different crop parameters.

Using sensors to measure differences in reflectance can be used to identify spatio-temporal variation within a field which may be linked to phenomena such as differences in chlorophyll content. These, in turn, may reflect crop nutritional status or availability of water, corresponding to differences in crop growth or biomass, or may indicate the presence of pests or diseases. Active sensors measuring light transmittance by red (650 nm) and NIR (940 nm) spectral bands correlate strongly with leaf chlorophyll content (N is a component of chlorophyll) and are used in targeted, variable rate N application, e.g. in applying fertilizer in response to plant deficiency during the vegetative stage (Solari et al., 2008).

One use of such reflectance data is vegetation indices (VIs) (Thenkabail et al., 2000; Hansen and Schjoerring, 2003). A VI is a single value based on comparing observations from multiple spectral bands. An example is the Simple Ratio (SR), which is the ratio between the reflectance recorded in the NIR and red bands: by distinguishing leaf greenness, it can be used to assess plant health or biomass. VIs frequently used to analyze crop characteristics include SR, Normalized Difference Vegetation Index (NDVI), Normalized Difference Red Edge (NDRE), Red Green Blue Vegetation Index, Soil Adjusted Vegetation Index, Infrared Percentage Vegetation Index, and Enhanced Vegetation Index (EVI and EVI2). The NDVI is a spectral index that relates NIR with red wavelength reflectance in a normalized form and has been shown to correlate well with plant biomass. In contrast to the NDVI, other indices relate to narrow wavelength ranges. One such index is the NDRE with the wavelengths 790 nm and 720 nm, which shows a strong correlation with N content in plants. VIs can be used to interpret sensor data in assessing plant growth and health.

4 Trade-offs in sensor data resolution

Spectral reflectance is one dimension of resolution, a measure of the level of detail captured in an image. There are four basic types of resolution:

- Spatial resolution: images have matrix structures composed of pixels: their smallest elements (Schrader and Pouncey, 1997). The spatial resolution of an image refers to the size that the pixel represents in the image. The smaller the pixel size, the higher the spatial resolution and the more detailed the image.
- Spectral resolution: refers to the number of bands (range of the EM spectrum) that a sensor can capture.
- Radiometric resolution: the level of gray that each pixel has, ranging from black to white. The gray level represents the intensity of the emitted/reflected EM energy measured by the sensor, corresponding to the pixel size.
- Temporal resolution: corresponding to when a satellite is able to return to the same area, i.e. the time required to collect a new image at the same point on the Earth's surface.

These are discussed in more detail later.

4.1 Spatial resolution

The spatial resolution of an image refers to the ground dimension represented by a pixel (representing a square area within an image, typically measured in square meters) (Fig. 4). The smaller the pixel size (and thus the area covered per pixel), the higher the image quality and resolution (Shiratsuchi, 2014; Paranhos Filho, 2021; Bullock, 2021). For example, the spatial resolution of wavelength bands 2, 3, 4 (visible), and 8 (infrared) of the Sentinel-2 satellite is 10 m, while bands 1, 9, and 10 (infrared) have a resolution of 60 m. Bands 5, 6, and 7 have a resolution of 20 m (Fletcher, 2012). Figure 5 shows different bands on the Sentinel-2 satellite for spatial resolution comparison. With the right-hand image shown in Fig. 5, we can observe a greater level of detail due to the smaller pixel size, including variations between and within fields.

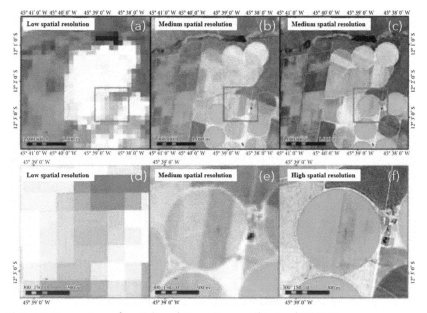

Figure 4 Comparison of spatial resolutions. Source: Shiratsuchi (2014).

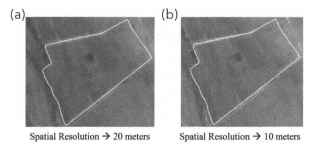

Figure 5 Comparison of Sentinel-2 bands resolutions between the red edge (a) and the near-infrared (b) waves.

4.2 Spectral resolution

Spectral resolution measures the sensitivity of a sensor in distinguishing differences in EM spectra (Paranhos Filho, 2021). For example, a sensor working in the range of 0.4–0.5 μm (micrometers) has a higher resolution compared to a sensor working in the range of 0.4–0.6 μm, meaning the first sensor will be able to register smaller variations in the EM spectrum (Novo, 2010). Figure 6 shows sensors with different spectral resolutions.

Ideally, remote imaging sensors need to capture images with multiple bands. Spectral resolution comprises three measurement parameters (Meneses and Almeida, 2012):

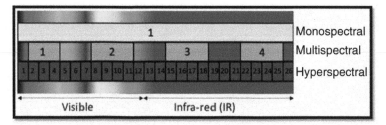

Figure 6 Different spectral resolutions. Source: Lira (2016).

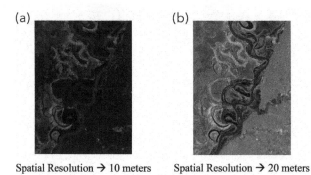

Figure 7 Comparison between band 3 green (10 m resolution) with band 6 red edge with (20 m resolution), Sentinel-2.

- Number of bands;
- The wavelength width of bands; and
- The positions of bands in the EM spectrum.

Sensors have a better spectral resolution if they have a greater number of bands located in different spectral regions and with a narrow wavelength. This greater number of spectral bands is necessary to differentiate a greater number of targets on the ground surface, e.g. targets that have a reflectance at a certain wavelength with a small difference from their surroundings.

Figure 7 shows the effect of spectral resolution in detecting different objects in a forest area. Even with a higher spatial resolution, image (a) does not show the river despite its size. This is because the reflectance of water and vegetation is practically the same at this wavelength, and thus, there is no contrast between the targets (Meneses and Almeida, 2012). Image (b) uses another band where the river can be easily identified: even with low reflectance the water can be detected and differentiated from the rest of the forest, establishing a contrast ratio between the targets observed.

Hyperspectral sensors have particularly high resolutions with bands measured in nanometers. An example is the Airborne Visible/Infrared

Imaging Spectrometer developed by the National Aeronautics and Space Administration, which is capable of acquiring continuous or practically continuous spectra along the reflected solar spectrum (0.4 μm to 2.5 μm) reaching 224 bands (Carvalho Júnior et al., 2003). Another example is the Moderate Resolution Imaging Spectroradiometer sensor that has 36 bands in the spectral ranges of near and visible infrared (400 nm to 2500 nm) and thermal infrared (8000 nm to 15 000 nm). In 2000, the Earth Observing-1 satellite was launched using the Hyperion sensor with 220 bands (Pizarro et al., 2001).

4.3 Radiometric resolution

Radiometric resolution refers to the number of levels of gray (ranging from black to white) the pixel can use to represent the information acquired by the sensor. The gray level represents the intensity of EM energy reflected or emitted by the sensor for the observed terrestrial surface area. The number of gray levels is measured in bits (Lira, 2016). The higher the gray scale, the greater the radiometric resolution and the higher the sensor's ability to differentiate small variations in the intensity of the reflected signal (Barbosa et al., 2019; Jensen, 2009). For example, the Landsat-5 satellite carries the Thematic Mapper sensor with an 8-bit radiometric resolution that corresponded to 256 gray levels (28 = 256 gray levels), while the Landsat-8 satellite equipped with the OLI sensor has a 12-bit resolution (128 = 4096 gray levels) (Barbosa et al., 2019). Figure 8 shows that an 8-bit image (28 = 256 gray levels) has better visual details compared to the lower radiometric resolution images of 6, 4, and 2 bits. The 2-bit image has only 4 levels of gray.

4.4 Temporal resolution

Sensors mounted on fixed structures can capture video images every few seconds. In RS, it takes time for a satellite to complete an orbit around the Earth (Lira, 2016) (Fig. 9). Satellite temporal resolution indicates the frequency with which the satellite revisits images in a particular area of interest; it is also known as the periodicity or repetitiveness with which the satellite records images

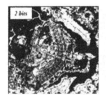 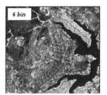

Figure 8 Band radiometric resolution in bits. Source: Meneses and Almeida (2012).

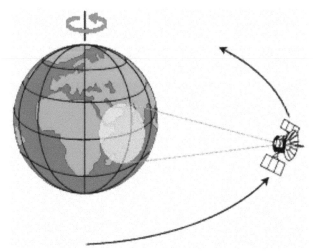

Figure 9 Path of a satellite around the Earth. Source: CCRS (1998).

(CCRS, 2019). The CBERS 4 satellite, e.g. has a temporal resolution of 26 days which limits its value for agricultural applications. In contrast, the Landsat-8 satellite has a temporal resolution of 16 days (USGS, 2018). The Sentinel series of satellites (some of which consist of two polar-orbiting satellites) has a temporal resolution of 10 days or 5 days where two satellites are paired (Van Der Meer et al., 2014; ESA, 2012). The Suomi-NPP platform provides daily imaging due to its wide imaging range (Martins, 2019).

4.5 Trade-offs in sensor data resolution

All sensors involve some degree of compromise between spatial and temporal resolution, and specificity. It is challenging in RS, e.g. to combine many spectral bands, high spectral resolution, and high spatial resolution, given the amount of data that needs to be collected. Normally high spatial resolution sensors have limited visible and NIR bands in order to reduce the large volume of data transmitted back to Earth. Some sensors with high spatial resolution and low spectral resolution (Meneses and Almeida, 2012). A sensor like the ASTER has a low spatial resolution that is compensated by a greater number of bands (14 in total).

5 Processing sensor data: sources of error and their resolution

Data to support PA and other applications is derived from a growing number of sources, including different types of terrestrial, aerial, and orbital sensor combinations. If correctly interpreted, this data can be used to better understand variations in crop response to fertilizer application (Basso et al., 2016) and to develop models to better target crop fertilization (Alesso et al., 2019). Sensors can also be used to detect patterns of crop biotic and abiotic stress (Bhandari et al., 2018) or to predict potential crop yields (Elavarasan et al., 2018; Peerlinck et al., 2018; El-Hendawy et al., 2019). However, this technology has several inherent sources of error which can cause incorrect spatial application rates. Sources of error include the accuracy of application maps, as well as errors in navigation and variable rate application (Chan et al., 2004).

The increasing volume and variety of data from a range of sensors and sensor platforms also bring with it challenges such as corrupting or 'noisy' data (data that contain a large amount of meaningless information) (Fan et al., 2014). Noisy data amass in proportion to the volume and variety of data collected. Working with spatial data presents additional challenges. A particular problem is spatial autocorrelation which describes the degree to which observations at different spatial locations resemble one another. Models based on data rely on observations being independent of one another. Spatial autocorrelation related to the geographic proximity of observations can undermine model reliability, e.g. through underestimation of degrees of model error. Additional data processing is required to account for how the location of the sample point influences its value and to avoid spatial autocorrelation effects when using spatial data in modeling (Ruß and Brenning, 2010). Effective methods to process 'big data' in this and other ways are critical to ensuring it can be translated into accurate and meaningful information for decision-making (McMillen, 2010; Chi et al., 2016; Kamilaris et al., 2017).

5.1 Sensor calibration

Calibration is the process of determining the relationship between the sensor signal and the characteristic(s) being measured in a given target. Calibration methods include empirical and mechanistic techniques (Weiss et al., 2020). A common empirical technique to calibrate sensors measuring VIs, such as NDVI, is the use of an N-rich strip area. This area has an adequate or slightly excess supply of a given input and is then used to generate a sensor signal value corresponding to 100% sufficiency for the variable being evaluated. Subsequent measurements taken of areas having different, and possibly insufficient, input amounts are then compared to the value of the reference area to calculate the sufficiency index of each sample area. This technique has been used to calibrate sensors used to measure N sufficiency in various crops using both active and passive sensors (Blackmer and Schepers, 1995; Holland and Schepers, 2013).

Calibration is particularly important in using passive sensors to create N variable-rate prescription maps. Small variations in sensor N readings can be masked by factors such as time of day, soil moisture, crop physiology, water stress, cloud coverage, shadow, etc. Figure 10 shows how UAS-collected reflectance data can be skewed compared to data from an ACS (Teixeira et al., 2020). If a passive sensor is used, it should be calibrated against data from active sensing technologies (Fig. 11). The use of active sensors in calibration is important, e.g. in estimating leaf area index (Viña et al., 2011) (Fig. 12).

5.2 Data filtering

Effective data filtering can reduce erroneous values in output datasets. Filtering techniques are generally performed to smooth an overall dataset by removing extreme and spatially nonstandard values (Blackmore, 1999). Common statistical data filtering methods can be used, such as removing values falling outside of a given range. This can, however, result in the removal of valid observations, so other methods of handling outlier values should be considered, such as Winsorization or a robust estimation method (Kwak and Kim, 2017).

It is important to account for the spatial nature of the data when considering filtering techniques. This is largely due to Tobler's first law of geography, which states, 'everything is related to everything else, but near things are more related than distant things.' One popular technique is to use

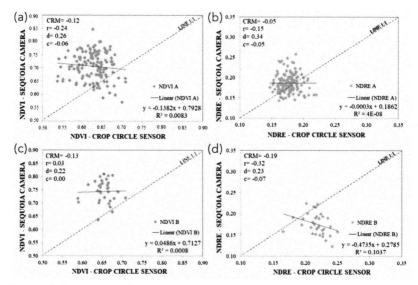

Figure 10 Comparison of active and passive sensors. Source: Adapted from Teixeira et al. (2020).

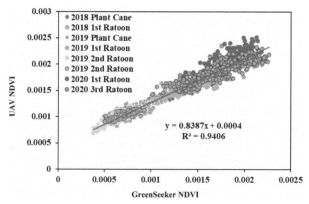

Figure 11 Correlation between active canopy sensors and unmanned aircraft systems. Source: Adapted from Forestieri (2021).

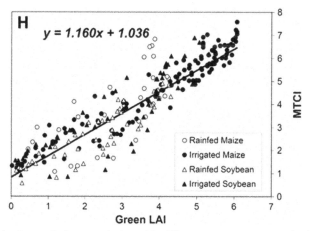

Figure 12 Leaf area index correlation with different crop systems and phenological stages. Source: Adapted from Viña et al. (2011).

a moving window-type calculation of the coefficient of variation (CV). Points falling within a given radius of the data point are used to calculate CV, which is then stored. This is done for every data point in the data set. Data points having a CV exceeding a given threshold are then discarded. This has the effect of smoothing the data by eliminating areas of abnormally high and low values (Spekken, 2013). Another technique is to remove data points occurring within a given distance of the field border since field borders are often subject to nonstandard growing conditions. In developing yield maps, semi-automated data filtering software has also been developed to facilitate the removal of erroneous data points, such as the USDA's Yield Editor software (Sudduth and Drummond, 2007).

5.3 Data analysis and evaluation

There have been significant developments in artificial intelligence and its use to analyze very large and complex datasets. Ruan et al. (2022) have used machine learning (ML) algorithms to analyze proximal sensing data from field experiments from multiple locations and years, combined with publicly available weather data, in order to predict winter wheat yields. ML algorithms have proved very promising for analyzing large volumes of sensor data. Some examples of ML algorithms are K-Nearest Neighbor (KNN), Multiple Linear Regression, Artificial Neural Networks – Multilayer Perceptron, and Random Forest (Morelli-Ferreira et al., 2021).

Researchers such as Fue et al. (2018), Xu et al. (2018), and Yeom et al. (2018) have used deep learning to identify regions of interest in RS images of cotton fields. Xu et al. (2018) developed a convolutional neural network to detect and count the number of newly opened cotton flowers in aerial photos as well as an algorithm to identify and calculate bolls in order to predict cotton yield. Tedesco-Oliveira et al. (2020) have used similar methods to rapidly estimate cotton yield via the identification and counting of bolls obtained from smartphone cameras taking pictures in fields. Morelli-Ferreira et al. (2021) have evaluated Sentinel satellite images for cotton yield prediction using eight VIs and four ML algorithms and concluded that the SR with the KNN algorithm improved the prediction of the timing of harvest and yield. Based on data from proximal active crop canopy sensors, Marchi (2021) used two VIs to identify variations in N requirements in cotton plants. They developed an algorithm for variable rate top dressing of cotton plants with N to optimize N use efficiency (NUE) and minimize N losses to the environment.

Habyarimana and Baloch (2021) have used Bayesian-based ML algorithms to correlate different VIs: fraction of absorbed photosynthetically active radiation (fAPAR) derived from Sentinel-2, chlorophyll content, and NDVI derived from handheld optical reflectance sensors. These techniques and data were used to develop and compare models to more accurately predict sorghum biomass yields. Modeling based on optical reflectance meter data measuring chlorophyll concentration index in the red edge at 700 nm was found to outperform NDVI meter-based models and was comparable to a model based on Sentinel-2-derived fAPAR.

6   Integrating remote and proximal sensor data for precision agriculture

Given issues such as the inherent complexity of the crop systems being studied, the differing strengths and weaknesses of different sensor systems, trade-offs in data resolution and quality, as well as potential sources of error from any data source, it makes sense to combine data from remote and proximal sensors. Multi-sensor fusion thus seeks to offset potential omissions and errors from one type of sensor data as well as minimize the uncertainty resulting from any estimated variable used (Barbedo, 2022). Figure 13 shows the integration of remote and proximal sensor data into a precision crop management system (Schellberg et al., 2008; Atzberger, 2013). An example is 'ground truthing.' The complexity of natural surfaces, the effects of atmosphere and the complex of relationship of spectral signatures to crop characteristics can limit the reliability of RS data which then needs to be validated by comparison with data from aerial or ground-based sensors (Steven, 2003). As an example, Hegarty-Craver et al. (2020) used images from UAVs to create a ground-truthing dataset which was then used to train a model capable of analyzing data from the Sentinel-1 and Sentinel-2 satellites in more accurately mapping maize crops. Techniques to combine data (data fusion) from the ground, aerial, and satellite sensors to identify site-specific variables such as N availability are discussed by Ahmad et al. (2022). The following discusses some examples of data fusion initiatives.

Pantazi et al. (2015) have combined data from satellite images, combine-harvest-mounted yield sensors and mobile proximal NIR soil sensors, using self-organizing clustering statistical techniques to create more accurate management zones for SSM. Benedetto et al. (2013) have used proximal EMI sensors, vegetative indices, and radiance data derived from satellite images, processed

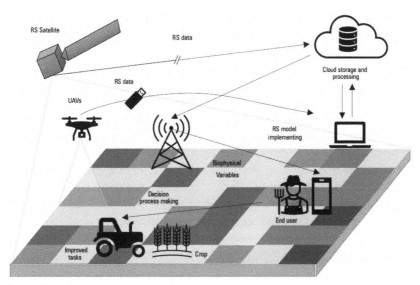

Figure 13 Collection of remote and proximal sensors data in agriculture. Source: Erazo-Mesa et al. (2022).

using multivariate geostatistics and clustering, to improve site-specific irrigation. More accurate yield maps have been developed using combinations of remote and proximal sensing data, crop modeling, and ML (Arab et al., 2021; Canata et al., 2021; Khanal et al., 2021) and by developing machine-vision-based yield monitor technologies (Jacques et al., 2021). Drones and other platforms carrying a range of sensors (hyperspectral, multispectral, and red-green-blue) have been used in areas such as weed control (Roslim et al., 2021). Tefera et al. (2022) used a combination of ground-based active sensors (Crop Circle) and aerial-based passive sensors (carried by UAS) to establish an optimal time for field pea establishment to outcompete the emergence of weeds in the field.

Dado et al. (2020) have compiled a ground-truth dataset of high-resolution (5 m) data from yield maps (derived from combine harvester data) from the US Midwest over the period 2008-18, representing over one million observations. This dataset was used to test the reliability of models derived from Landsat and Sentinel-2 sensor data and suggest ways these might be refined to better predict soybean yields. In better exploiting existing datasets both to develop and validate models, Woodard (2018) has developed the AgAnalytics platform for researchers to share publicly available data more efficiently by improving the ways it is documented, disseminated, and processed.

Maestrini and Basso (2018) analyzed yield monitor data (involving 1625 yield maps) collected by proximal combine-mounted sensors for 338 fields growing maize, soybean, cotton, and wheat across 8 states in the US Midwest. These data were used to identify patterns of in-field spatial and temporal variation. These data were correlated with publicly available datasets relating to soil type, topography, and rainfall patterns to explain yield fluctuations in the same locations in fields (characterized as unstable) which the authors were able to relate to the interaction of topography and rainfall. A Landsat-derived NDVI (red band at 0.3 m resolution and temperature at 2 m resolution) was found to be helpful in the timing of N application in these unstable field zones to improve NUE and yields.

Current data fusion techniques are reviewed by Barbedo (2022). While they have improved accuracy in some areas (e.g. improving image identification), applications to more complex problems such as improving the accuracy of management zones have been mixed. This is, in part, due to continuing challenges in correlating different data sources and due to the fact that, though large amounts of data are being generated, they are often still not sufficient to cover all the variables in complex agricultural environments.

7 Conclusion

The use of data from sensors has been critical in developing effective PA applications. Research suggests that site-specific variable rate N application, e.g. can increase NUE and reduce N leaching losses without compromising yield, with some research suggesting potential savings of 2.25–4.5 kg N/hectare (Long, 2019; Pedersen et al., 2019). A recent study by the US Association of Machinery Manufacturers in alliance with the American Soybean Association, the National Corn Growers Association, and CropLife America suggested that PA techniques were responsible for a 7% improvement in fertilizer placement efficiency, a 9% reduction in herbicide, and 4% reduction in water use with the potential for far greater improvements (https://www.aem.org/news/the-environmental-benefits-of-precision-agriculture-quantified).

However, the process of collecting and processing data to deliver actionable outcomes remains complex, with a significant potential for error in the interpretation of sensor data to achieve more sustainable crop management. Understanding and avoiding such errors requires an understanding of the key elements in this process, the limitation of sensor information, and what different sensors are able to detect or measure, as well as parameters such as spatial, spectral, and temporal resolution that meet the spatial and temporal needs of management, i.e. match the spatial and temporal variation of the target property. It includes an understanding of different types of sensor as well as the different platforms on which sensors can be mounted (terrestrial/proximal, aerial, and orbital), their relative strengths and weaknesses, and the ways they can complement each other. In general terms, RS can be seen to be best suited to identifying where variability occurs and where and when an intervention might be targeted, while proximal sensing can provide more precise information on why variability has occurred and what treatment and rate of application might be needed. When combined, the two can achieve the goal of PA in applying the right product or treatment at the right location, rate, and time.

8 References

Adao, T., Hruška, J., Pádua, L., Bessa, J., Peres, E., Morais, R. and Sousa, J. (2017). Hyperspectral imaging: A review on UAV-based sensors, data processing and applications for agriculture and forestry. *Remote Sens.* 9(11), 1110. doi: 10.3390/rs9111110.

Ahmad, U., Nasirahmadi, A., Hensel, O. and Marino, S. (2022). Technology and data fusion models to enhance site-specific crop monitoring. *Agronomy* 12(3), 555.

Alesso, C. A., Cipriotti, P. A., Bollero, G. A. and Martin, N. F. (2019). Experimental designs and estimation methods for on-farm research: A simulation study of corn yields at field scale. *Agron. J.* 111(6), 2724-2735. doi: 10.2134/agronj2019.03.0142.

Arab, S. T., Noguchi, R., Matsushita, S. and Ahamed, T. (2021). Prediction of grape yields from time-series vegetation indices using satellite remote sensing and a machine-learning approach. *Remote Sens. Appl. Soc. Environ.* 22, 100485. doi: 10.1016/j.rsase.2021.100485.

Atzberger, C. (2013). Advances in remote sensing of agriculture: Context description, existing operational monitoring systems and major information needs. *Remote Sens.* 5(2), 949-981. doi: 10.3390/rs5020949.

Barbedo, J. G. A. (2022). Data fusion in agriculture: Resolving ambiguities and closing data gaps. *Sensors (Basel)* 22(6), 2285.

Barbosa, C. C. F., Novo, E. M. L. M. and Martins, V. S. (2019). *Introdução ao sensoriamento remoto de sitemas aquáticos: Princípios e aplicações.* São José dos Campos, Brazil: INPE. Available at: http://www.dpi.inpe.br/labisa/livro/res/conteudo.pdf (accessed 28 February 2022).

Basso, B., Fiorentino, C., Cammarano, D. and Schulthess, U. (2016). Variable-rate nitrogen fertilizer response in wheat using remote sensing. *Precis. Agric.* 17(2), 168-182. doi: 10.1007/s11119-015-9414-9.

Benedetto, D., Castrignano, A., Diacono, M., Rinaldi, M., Ruggieri, S. and Tamborrino, R. (2013). Field partition by proximal and remote sensing data fusion. *Biosyst. Eng.* 114(4), 372-383. doi: 10.1016/j.biosystemseng.2012.12.001.

Bhandari, S., Raheja, A., Chaichi, M. R., Green, R. L., Do, D., Pham, F. H., Ansari, M., Wolf, J. G., Sherman, T. M. and Espinas, A. (2018). Lessons learned from UAV-based remote sensing for precision agriculture. In: *International conference on unmanned aircraft systems (ICUAS)*, USA.

Blackmer, T. M. and Schepers, J. S. (1995). Use of a chlorophyll meter to monitor nitrogen status and schedule fertigation for corn. *J. Prod. Agric.* 8(1), 56-60. doi: 10.2134/jpa1995.0056.

Blackmore, S. (1999). Remedial correction of yield map data. *Precis. Agric.* 1(1), 53-66.

Brandão, Z. N. (2009). *Estimativa da produtividade e estado nutricional da cultura do algodão irrigado via técnicas de sensoriamento remoto. 152 f. Tese* (PhD student in Natural Resources), Universidade Federal de Campina Grande, Campina Grande, Brazil.

Bullock, D. S. (2021). Using satellite imagery to estimate optimal site-specific N side-dressing strategies [abstract]. In: *SSSA International Annual Meeting*, CSSA, ASA, Salt Lake City, UT. Available at: https://scisoc.confex.com/scisoc/2021am/meetingapp .cgi/Paper/132907.

Canata, T. F., Wei, M. C. F., Maldaner, L. F. and Molin, J. P. (2021). Sugarcane yield mapping using high-resolution imagery data and machine learning technique. *Remote Sens.* 13(2), 232. doi: 10.3390/rs13020232.

Carvalho Júnior, O. A., Guimarães, P. H., Lopes, R. A. S., Guimarães, R. F., Martins, É. D. S. and Pedreño, J. N. (2003). *Estratificação dos Ambientes de Mistura em Imagens Hiperespectrais.* Simpósio Brasileiro de Sensoriamento Remoto, Brazil, pp. 1045-1051.

CCRS: Canada Centre for Remote Sensing. (2019). Fundamentals of remote sensing. Government of Canada. Available at: https://www.nrcan.gc.ca/sites/www.nrcan .gc.ca/files/earthsciences/pdf/resource/tutor/fundam/pdf/fundamentals_e.pdf (accessed 19 February 2022).

Chan, C. W., Schueller, J. K., Miller, W. M., Whitney, J. D. and Cornell, J. A. (2004). Error sources affecting variable rate application of nitrogen fertilizer. *Precis. Agric.* 5(6), 601-616.

Chi, M., Plaza, A., Benediktsson, J. A., Sun, Z., Shen, J. and Zhu, Y. (2016). Big data for remote sensing: Challenges and opportunities. *Proc. IEEE* 104(11), 2207-2219. doi: 10.1109/JPROC.2016.2598228.

Colaço, A. F. and Bramley, R. G. V. (2018). Do crop sensors promote improved nitrogen management in grain crops? *Field Crops Res.* 218, 126–140. doi: 10.1016/j. fcr.2018.01.007.

Crepani, E. (1993). *Princípios Básicos de Sensoriamento Remote.* CNPQ/INPE, São José dos Campos, Brazil, p. 45.

Dado, W. T., Deines, J. M., Patel, R., Liang, S. and Lobell, D. B. (2020). High-resolution soybean yield mapping across the US midwest using subfield harvester data. *Remote Sens.* 12(21), 3471. doi: 10.3390/rs12213471.

Dash, J., Jeganathan, C. and Atkinson, P. M. (2010). The use of MERIS terrestrial chlorophyll index to study spatio-temporal variation in vegetation phenology over India. *Remote Sens. Environ.* 114(7), 1388–1402. doi: 10.1016/j.rse.2010.01.021.

Elavarasan, D., Vincent, D. R., Sharma, V., Zomaya, A. Y. and Srinivasan, K. (2018). Forecasting yield by integrating agrarian factors and machine learning models: A survey. *Comput. Electron. Agric.* 155, 257–282. doi: 10.1016/j.compag.2018.10.024.

Elbl, J., Mezera, J., Kintl, A., Širůček, P. and Lukas, V. (2021). Comparisons of uniform and variable rate nitrogen fertilizer applications in real conditions–Evaluation of potential impact on the yield of wheat available for use in animal feed. *Acta Univ. Agric. Silvic Mendelianae Brun* 69(1), 33–43.

El-Hendawy, S., Al-Suhaibani, N., Elsayed, S., Refay, Y., Alotaibi, M., Dewir, Y. H., Hassan, W. and Schmidhalter, U. (2019). Combining biophysical parameters, spectral indices and multivariate hyperspectral models for estimating yield and water productivity of spring wheat across different agronomic practices. *PLoS ONE* 14(3), e0212294. doi: 10.1371/journal.pone.0212294.

ESA: European Space Agency. (2012). Sentinel-2: ESA's optical high-resolution mission for GMES operational services. European Space Agency. Available at: https:// sentinel.esa.int/documents/247904/349490/s2_sp-1322_2.pdf (accessed 12 February 2022).

Fan, J., Han, F. and Liu, H. (2014). Challenges of Big Data analysis. *Natl. Sci. Rev.* 1(2), 293–314. doi: 10.1093/nsr/nwt032.

Ferguson, R. B. (2019). Proximal crop sensing. In: Stafford, J. (Ed.), *Precision Agriculture for Sustainability.* Burleigh Dodds Science Publishing, Cambridge, UK, pp. 3–28.

Fletcher, K. (2012). Sentinel-2: ESA's optical high-resolution mission for GMES operational services. *Esa SP-1322.* Available at: https://sentinel.esa.int/documents /247904/349490/s2_sp-1322_2.pdf (accessed 2 March 2012; accessed 15 February 2022).

Florenzano, T. G. (2011). *Iniciação em Sensoriamento Remoto* (3rd edn.). ampl. e atual, Oficina de Textos, Sao Paulo, Brazil, p. 128.

Foley, J. A., Ramankutty, N., Brauman, K. A., Cassidy, E. S., Gerber, J. S., Johnston, M., Mueller, N. D., O'Connell, C., Ray, D. K., West, P. C., Balzer, C., Bennett, E. M., Carpenter, S. R., Hill, J., Monfreda, C., Polasky, S., Rockström, J., Sheehan, J., Siebert, S., Tilman, D. and Zaks, D. P. (2011). Solutions for a cultivated planet. *Nature* 478(7369), 337–342. doi: 10.1038/nature10452.

Fue, K. G., Porter, W. M. and Rains, G. C. (2018). *Deep Learning-Based Real-Time GPU-Accelerated Tracking and Counting of Cotton Bolls Under Field Conditions Using a Moving Camera.* American Society of Agricultural and Biological Engineers. doi: 10.13031/aim.201800831.

Ge, X., Schaap, M., Kranenburg, R., Segers, A., Reinds, G. J., Kros, H. and de Vries, W. (2020). Modeling atmospheric ammonia using agricultural emissions with improved spatial variability and temporal dynamics. *Atmos. Chem. Phys.* 20(24), 16055–16087. doi: 10.5194/acp-20-16055-2020.

Gebbers, R. (2019). Proximal soil surveying and monitoring techniques. In: Stafford, J. (Ed.), *Precision Agriculture for Sustainability*. Burleigh Dodds Science Publishing, Cambridge, UK, pp. 29–78.

Gitelson, A. A., Viña, A., Verma, S. B., Rundquist, D. C., Arkebauer, T. J., Keydan, G., Leavitt, B., Ciganda, V., Burba, G. G. and Suyker, A. E. (2006). Relationship between gross primary production and chlorophyll content in crops: Implications for the synoptic monitoring of vegetation productivity. *J. Geophys. Res.* 111(D8). doi: 10.1029/2005JD006017.

Guo, Q. H., Wu, F., Hu, T., Chen, L., Liu, J., Zhao, X., Gao, S. and Pang, S. (2016). Perspectives and prospects of unmanned aerial vehicle in remote sensing monitoring of biodiversity. *Biodivers. Sci.* 24(11), 1267–1278. doi: 10.17520/biods.2016105.

Habyarimana, E. and Baloch, F. S. (2021). Machine learning models based on remote and proximal sensing as potential methods for in-season biomass yields in commercial sorghum fields. *PLoS ONE* 16(3), e0249136. doi: 10.1371/journal.pone.0249136.

Hansen, P. M. and Schjoerring, J. K. (2003). Reflectance measurement of canopy biomass and nitrogen status in wheat crops using normalized difference vegetation indices and partial least squares regression. *Remote Sens. Environ.* 86(4), 542–553. doi: 10.1016/S0034-4257(03)00131-7.

Hegarty-Carver, M., et al. (2020). Remote crop mapping at scale: Using satellite images and unmanned aerial vehicle-acquired data as ground truth. *Remote Sens.* 12: 1984.

Holland, K. H. and Schepers, J. S. (2013). Use of a virtual-reference concept to interpret active crop canopy sensor data. *Precis. Agric.* 14(1), 71–85. doi: 10.1007/s11119-012-9301-6.

Huang, X., Cao, Y. and Li, J. (2020). An automatic change detection method for monitoring newly constructed building areas using time-series multi-view high-resolution optical satellite images. *Remote Sens. Environ.* 244, 111802. doi: 10.1016/j.rse.2020.111802.

Hunt, L. and Daughtry, C. (2018). What good are unmanned aircraft systems for agricultural remote sensing in precision agriculture? *J. Remote Sens.* 39(15–16), 5345–5376.

ISPA. (2022). International society of precision agriculture. *Precision AG Definition*. Available at: https://www.ispag.org/about/definition (accessed 20 May 2022).

Jacques, A. A. B., Adamchuk, V. I., Park, J., Cloutier, G., Clark, J. J. and Miller, C. (2021). Towards a machine vision-based yield monitor for the counting and quality mapping of shallots. *Front. Robot. AI* 8, 41. doi: 10.3389/frobt.2021.627067.

Jensen, J. R. (1996). *Introductory Digital Image Processing: A Remote Sensing Perspective*. Prentice Hall, Upper Saddle River, p. 316.

Jensen, J. R. (2009). *Sensoriamento Remoto do Ambiente: Uma perspectiva Em Recursos Terrestres*. Parêntese, São José dos Campos, Brazil, p. 598.

Kamilaris, A., Kartakoullis, A. and Prenafeta-Boldú, F. X. (2017). A review on the practice of big data analysis in agriculture. *Comput. Electron. Agric.* 143, 23–37. doi: 10.1016/j.compag.2017.09.037.

Khanal, S., Klopfenstein, A., Kc, K., Ramarao, V., Fulton, J., Douridas, N. and Shearer, S. A. (2021). Assessing the impact of agricultural field traffic on corn grain yield using remote sensing and machine learning. *Soil Till. Res.* 208, 104880. doi: 10.1016/j. still.2020.104880.

Kumar, R. and Silva, L. (1973). Light ray tracing through a leaf cross section. *Appl. Opt.* 12(12), 2950-2954. doi: 10.1364/AO.12.002950.

Kwak, S. K. and Kim, J. H. (2017). Statistical data preparation: Management of missing values and outliers. *Korean J. Anesthesiol.* 70(4), 407-411. doi: 10.4097/ kjae.2017.70.4.407.

Lichtenthaler, H. K. and Buschmann, C. (2001). *Chlorophylls and Carotenoids: Measurement and Characterization by UV-Vis Spectroscopy, Current Protocols in Food Analytical Chemistry*. John Wiley and Sons, New York, NY, pp. F4.3.1–F4.3.8.

Lira, C. (2016). Sistemas de informação geográfica: análise de dados de satélite, Lisboa, Portugal. Available at: https://www.researchgate.net/publication/312383752_ Sistemas_de_Informacao_Geografica_Analise_de_Dados_de_Satelite (accessed 26 March 2022).

Long, D. S. (2019). Site-specific nutrient management systems. In: Stafford, J. (Ed.), *Precision Agriculture for Sustainability*. Burleigh Dodds Science Publishing, Cambridge, UK, pp. 299-322.

Lorenzzetti, J. A. (2015). *Princípios Físicos de Sensoriamento Remote*. Blucher, São Paulo, Brazil, p. 293.

Maestrini, B. and Basso, B. (2018). Predicting spatial patterns of within-field crop yield variability. *Field Crops Res.* 219(1), 106–112. doi: 10.1016/j.fcr.2018.01.028.

Marchi, J. de (2021). Active Sensor Algorithm Approach to Optimize Nitrogen Rate Fertilization in Cotton Production. *Preprints*. doi: 10.20944/preprints202110.0183. v1.

Martins, V. S. (2019). *Introdução Ao Sensoriamento Remoto de Sistemas Aquáticos: Principios e Aplicações* (1st edn.). INPE, São José dos Campos, Brazil. Available at: http://www.dpi.inpe.br/labisa/livro/res/conteudo.pdf (accessed 22 February 2022).

McMillen, D. P. (2010). Issues in spatial data analysis. *J. Reg. Sci.* 50(1), 119-141. doi: 10.1111/j.1467-9787.2009.00656.x.

Meneses, P. R. and Alemida, T. de (2012). Introdução ao Processamento de Imagens em Sensoriamento Remoto, Brasília, Brazil. Available at: https://docero.com.br/doc/ ss1x1x5 (accessed 23 March 2022).

Mobley, C. D. (1994). *Light and Water: Radiative Transfer in Natural Waters*. Academic Press, San Diego, p. 592.

Molin, J. P., Amaral, L. R. and Colaço, A. F. (2015). *Agricultura de Precisão*. Oficina de Textos, São Paulo, pp. 119-153.

Moreira, M. A. (2011). *Fundamentos de Sensoriamento Remoto e Metodologias de Aplicação*. 4 Universidade Federal de Viçosa (UFV), Viçosa, Brazil, p. 422.

Morelli-Ferreira, F., Maia, N. J. C., Tedesco, D., Kazama, E. H., Carneiro, F. M., Santos, L. B., Junior, G. F. S., Rolim, G. S., Shiratsuchi, L. S. and Silva, R. P. (2021). Comparison of machine learning techniques in cotton yield prediction using satellite remote sensing. *Preprints* 1, 1-17. doi: 10.20944/preprints202112.0138.v2.

Mulla, D. J. (2013). Twenty-five years of remote sensing in precision agriculture: Key advances and remaining knowledge gaps. *Biosystems Engineering* 114(4), 358-371.

Narvaez, F. Y., Reina, G., Torres-Torriti, M., Kantor, G. and Cheein, F. A. (2017). A survey of ranging and imaging techniques for precision agriculture phenotyping. *IEEE ASME Trans. Mechatron.* 22(6), 2428-2439. doi: 10.1109/TMECH.2017.2760866.

Novo, E. M. L. M. (2008). *Sensoriamento Remoto: Princípios e Aplicações*. Blucher, São Paulo, Brazil, p. 363.

Novo, E. M. L. de M. (2010). Sensoriamento remoto. In: *Princípios e aplicações* (Vol. 4). Blucher, São Paulo, Brazil. Available at: https://www.atenaeditora.com.br/wp-content /uploads/2018/10/E-book-Aplica%C3%A7%C3%B5es-e-Princ%C3%ADpios-do -Sensoriamento-Remoto-1.pdf (accessed 10 March 2022).

Pantazi, X., Moshou, D., Mouazen, A. M., Alexandridis, T. and Kuang, B. (2015). Data fusion of proximal sensing and remote sensing for the delineation of management zones in arable precision farming. In: Anon (Ed.), *Proceedings of the Seventh International Conference of Information and Communication Technologies in Agriculture, Food and the Environment*, Kavala, Greece.

Paranhos Filho, A. C. (2021). Chapter 3: Resolução de imagens de satelite. In: Filho, P. (Ed.), *Geotecnologia para Aplicaçao Ambiental*. PR Uniedusul, Maringá, Brazil, pp. 25-59.

Pedersen, S. M., Medici, M., Anken, T., Tohidloo, G., Pedersen, M. F., Carli, G., Canavari, M., Tsiropoulos, Z. and Fountas, S. (2019). Financial and environmental performance of integrated precision farming systems. In: Stafford, J. (Ed.), *Agriculture*. Wageningen Academic Publishers, Wageningen, The Netherlands, pp. 833-839.

Peerlinck, A., Sheppard, J. and Maxwell, B. (2018). Using deep learning in yield and protein prediction of winter wheat based on fertilization prescriptions in precision agriculture. In: *International Conference on Precision Agriculture (ICPA)*, Canada.

Pizarro, M. A., Epiphanio, J. C. N. and Galvão, L. S. (2001). Caracterização mineralógica de solos tropicais por sensoriamento remoto hiperespectral. *Pesq. Agropec. Bras.* 36(10), 1277-1286. Available at: https://www.scielo.br/j/pab/a/Wqsh4D8XyKN5JGy fwJBC7mc/?lang=pt (accessed 14 February 2022).

Queiroz, D. Md, Coelho, A. LdF., Valente, D. S. M. and Schueller, J. K. (2020). Sensors applied to digital agriculture: A review. *Rev. Cienc. Agron.* 51(5), e20207751. doi: 10.5935/1806-6690.20200086.

Roslim, M. H. M., Juraimi, A. S., Che'Ya, N. N., Sulaiman, N., Manaf, M. N. H. A., Ramli, Z. and Motmainna, M. (2021). Using remote sensing and an unmanned aerial system for weed management in agricultural crops: A review. *Agronomy* 11(9), 1809. doi: 10.3390/agronomy11091809.

Ruan, G., Li, X., Yuan, F., Cammarano, D., Ata-Ul-Karim, S. T., Liu, X., Tian, Y., Zhu, Y., Cao, W. and Cao, Q. (2022). Improving wheat yield prediction integrating proximal sensing and weather data with machine learning. *Comput. Electron. Agric.* 195(1), 106852. doi: 10.1016/j.compag.2022.106852.

Ruß, G. and Brenning, A. (2010). Data mining in precision agriculture: Management of spatial information. In: *International Conference on Information Processing and Management of Uncertainty in Knowledge-Based Systems*. Springer, pp. 350-359. Available at: https://link.springer.com/chapter/10.1007/978-3-642-14049-5_36 (accessed 28 May 2022).

Santos, A. F., Corrêa, L. N., Lacerda, L. N., Tedesco-Oliveira, D., Pilon, C., Vellidis, G. and da Silva, R. P. (2021). High-resolution satellite image to predict peanut maturity variability in commercial fields. *Precis. Agric.* 22(5), 1464-1478. doi: 10.1007/ s11119-021-09791-1.

Schellberg, J., Hill, M. J., Gerhards, R., Rothmund, M. and Braun, M. (2008). Precision agriculture on grassland: Applications, perspectives and constraints. *Eur. J. Agron.* 29(2-3), 59-71. doi: 10.1016/j.eja.2008.05.005.

Schielack, V. and Thomassen, J. (2016). Harvester-based sensing system for cotton fiber quality. *J. Cotton Sci.* 20, 393.

Schrader, S. and Pouncey, R. (1997). *Erdas Field Guide. 4.* Erdas Inc., Atlanta, USA, p. 656.

Shiratsuchi, L. S. (2014). Sensoriamento Remoto: conceitos básicos e aplicações na Agricultura de Precisão. In: Bernardi, A. C. C. (Ed.), *Agricultura de Precisão: Resultados de um Novo Olhar* (Vol. 4). EMBRAPA, Brasília, Brazil, pp. 58–73.

Singh, G., Reynolds, C., Byrne, M. and Rosman, B. (2020). A remote sensing method to monitor water, aquatic vegetation, and invasive water hyacinth at national extents. *Remote Sens.* 12(24), 4021. doi: 10.3390/rs12244021.

Solari, F., Shanahan, J., Ferguson, R., Schepers, J. and Gitelson, A. (2008). Active sensor reflectance measurements of corn nitrogen status and yield potential. *Agron. J.* 100(3), 571–579. doi: 10.2134/agronj2007.0244.

Spekken, M. (2013). A simple method for filtering spatial data. *Precision Agriculture* 13, 259–266. Springer. Available at: https://www.agriculturadeprecisao.org.br/wp-content/uploads/2019/11/cgr-2013_07.pdf (accessed 28 May 2022).

Steven, M. D. (2003). Ground truth: An underview. *Int. J. Remote Sens.* 8(7), 1033–1038.

Sudduth, K. A. and Drummond, S. T. (2007). Yield editor: Software for removing errors from crop yield maps. *Agron. J.* 99(6), 1471–1482. doi: 10.2134/agronj2006.0326.

Sui, R., Thomasson, J. A. and Ge, Y. (2012). Development of sensor systems for precision agriculture in cotton. *Int. J. Agric. Biol. Eng.* 5(4), 1–14.

Tedesco, D., de Oliveira, M. F., dos Santos, A. F., Silva, E. H. C., de Souza Rolim, G. and da Silva, R. P. (2021). Use of remote sensing to characterize the phenological development and to predict sweet potato yield in two growing seasons. *Eur. J. Agron.* 129, 126337. doi: 10.1016/j.eja.2021.126337.

Tedesco-Oliveira, D., Pereira da Silva, R., Maldonado, W. and Zerbato, C. (2020). Convolutional neural networks in predicting cotton yield from images of commercial fields. *Comput. Electron. Agric.* 171(1), 105307. doi: 10.1016/j.compag.2020.105307.

Tefera, A. T., Banerjee, B. P., Pandey, B. R., James, L., Puri, R. R., Cooray, O., Marsh, J., Richards, M., Kant, S., Fitzgerald, G. J. and Rosewarne, G. M. (2022). Estimating early season growth and biomass of field pea for selection of divergent ideotypes using proximal sensing. *Field Crops Res.* 277, 108407. doi: 10.1016/j.fcr.2021.108407.

Teixeira, A. A. D., Mendes Júnior, C. W., Bredemeier, C., Negreiros, M. and Aquino, RdS. (2020). Evaluation of the radiometric accuracy of images obtained by a sequoia multispectral camera. *Eng. Agric.* 40(6), 759–768. doi: 10.1590/1809-4430-eng.agric.v40n6p759-768/2020.

The Royal Society. (2016). *Reaping the Benefits: Science and the Sustainable Intensification of Global Agriculture.* Available at: http://royalsociety.org/Reapingthebenefits (accessed 12 December 2016).

Thenkabail, P. S., Smith, R. B. and De Pauw, E. (2000). Hyperspectral vegetation indices and their relationships with agricultural crop characteristics. *Remote Sens. Environ.* 71(2), 158–182. doi: 10.1016/S0034-4257(99)00067-X.

USGS: United States Geological Survey. (2018). Landsat 8 OLI (operational land Imager) and TIRS (thermal infrared sensor). Available at: https://lta.cr.usgs.gov/L8 (accessed 28 March 2022).

Van Der Meer, F. D., Van Der Werff, H. M. A. and Van Ruitenbeek, F. J. A. (2014). Potential of ESA's Sentinel-2 for geological applications. *Remote Sens. Environ.* 148, 124–133. doi: 10.1016/j.rse.2014.03.022.

Viña, A., Gitelson, A. A., Nguy-Robertson, A. L. and Peng, Y. (2011). Comparison of different vegetation indices for the remote assessment of green leaf area index of crops. *Remote Sens. Environ.* 115(12), 3468–3478. doi: 10.1016/j.rse.2011.08.010.

Viscarra Rossel, R. A., McKenzie, N. J., Adamchuk, V. I., Sudduth, K. A. and Lobsey, C. (2011). Proximal soil sensing: An effective approach for soil measurements in space and time. *Adv. Agron.* 113, 237–282.

Weiss, M., Jacob, F. and Duveiller, G. (2020). Remote sensing for agricultural applications: A meta-review. *Remote Sens. Environ.* 236, 111402, doi: 10.1016/j.rse.2019.111402.

Woodard, J. (2018). *Aganalytics: Webplatform*. Personal Communication. Available at: http://analytics.ag.

Xu, R., Li, C., Paterson, A. H., Jiang, Y., Sun, S. and Robertson, J. S. (2018). Aerial images and convolutional neural network for cotton bloom detection. *Front. Plant Sci.* 8(1), 17. doi: 10.3389/fpls.2017.02235.

Yang, C. (2019). Airborne and satellite remote sensors for precision agriculture. In: Stafford, J. (Ed.), *Precision Agriculture for Sustainability*. Burleigh Dodds Science Publishing, Cambridge, UK.

Yeom, J., Jung, J., Chang, A., Maeda, M. and Landivar, J. (2018). Automated open cotton bol detection for yield estimation using unmanned aircraft vehicle (UAV) data. *Remote Sens.* 10(12), 1895. doi: 10.3390/rs10121895.

Yuan, G., Wang, Y., Zhao, F., Wang, T., Zhang, L., Hao, M., Yan, S., Dang, L. and Peng, B. (2021). Accuracy assessment and scale effect investigation of UAV thermography for underground coal fire surface temperature monitoring. *Int. J. Appl. Earth Obs. Geoinf.* 102, 102426. doi: 10.1016/j.jag.2021.102426.

Zanotta, D. C., Ferreira, M. P. and Zortea, M. (2019). Chapter 1: Iniciação aos dados de sensoriamento remoto. In: Zanotta, D. C., Ferreira, M. P. and Zortea, M. (Eds.), *Processamento de Imagens de Satélite*. Oficina de Texto, São Paulo, Brazil, pp. 11–63.

Zhang, C. and Kovacs, J. M. (2019). The use of unmanned aerial systems (UASs) in precision agriculture. In: Stafford, J. (Ed.), *Precision Agriculture for Sustainability*. Burleigh Dodds Science Publishing, Cambridge, UK.

Zheng, H., Zhou, X., He, J., Yao, X., Cheng, T., Zhu, Y., Cao, W. and Tian, Y. (2020). Early season detection of rice plants using RGB, NIR-G-B and multispectral images from unmanned aerial vehicle (UAV). *Comput. Electron. Agric.* 169, 105223. doi: 10.1016/j.compag.2020.105223.

Ingram Content Group UK Ltd.
Milton Keynes UK
UKHW021259070623
423032UK00016B/95